Gravity Control

With Present Technology

Gravity Control

With Present Technology

Dr. Frederick Alzofon

Commentary by David Alzofon

This book includes sections that have been previously published in the article "Frederick E. Alzofon, 'The Unity of Nature and the Search for a Unified Field Theory' ", Physics Essays, Vol. 6, pp. 599-608 (1993) (http://www.physicsessays.org/browse-journal-2/product/677-17-frederick-e-alzofon-the-unity-of-nature-and-the-search-for-a-unified-field-theory.html). They are reprinted with permission of Physics Essays Publication.

ISBN-13: 978-1548293154

ISBN:-10: 1548293156

Createspace Independent Publishing Platform, North Charleston, SC

A Klaf Raknar MediaWorks Production

For my Dad

.

.

ACKNOWLEDGMENTS

No book is an island.

First and foremost, I thank my father for inspiring me with his love of science, mathematics, and Western culture. But his greatest gift was hope. Shakespeare said, "The miserable have no other medicine, but only hope." My father, on the other hand, gave hope a name and backed it with hard science. This was far too precious a gift to allow to die with him, and this book was written to ensure that it might be passed on to others in time to make a difference. My mom's role in supporting my dad must not be forgotten either.

I am deeply grateful for the help of Barbara Wolff, Einstein Information Officer, Albert Einstein Archives, The Hebrew University of Jerusalem, for providing photos of my father's 1955 letter to Dr. Einstein, which I thought had been lost to history. The letter was vitally important in establishing a timeline for my father's meeting with Dr. Feynman at Caltech and showing the evolution of his thought concerning gravitation.

Thanks to Dave Conklin for providing morale-boosting iconoclasm during the Stanford years and beyond.

Sincere gratitude to Dr. Han Kim for being a good friend, a great raconteur, and a true humanitarian.

Last, but far from least, I would like to thank my significant other, Susan O'Neill, who was a forever faithful and supportive friend and ally throughout twenty years of this saga. Susan knew my father well and witnessed every chapter of the Silicon Valley era investor search, only a fraction of which is mentioned here. Her advice and feedback was always thoughtful, intelligent, and valuable. Her patience, too, was plentiful. When I say that I'm glad this project is finally at an end, I know that she will more than second my emotion.

BOOK I and BOOK II

Overview

Book I

Gravity Control: Past, Present, and Future

Book II

Theoretical Foundation and Applied Technology

Postscripts

TABLE OF CONTENTS

BOOK I

Gravity Control: Past, Present, and Future

By David Alzofon

BOOK II

Theoretical Foundation and Applied Technology

By Dr. Frederick Alzofon, Commentary by David Alzofon

AUTHOR BIOGRAPHY *271*

BOOK I

Gravity Control: Past, Present, and Future

by David Alzofon

Part I

A Transformational Technology

Chapter 0

The Time Machine

THE BOOK you are holding is the last link in a chain of events dating back to 1942, when my father, age 23, entered the PhD program in Physics at Cal Berkeley. It may also be the *final* link. And strange to say, *you* could have a hand in deciding the outcome of this saga.

As I understand it from my mom, as well as my aunt and uncle, all of whom worked on the Manhattan Project, my dad was a bit of a troublemaker, perpetually questioning fundamental assumptions when other students were copying the equations. And when his professors *did* get down to the equations, he was not shy about correcting them when he saw an error go up on the blackboard. None of this endeared him to J. Robert Oppenheimer, Raymond T. Birge, and a few others. He did, however, form lasting friendships with Victor Lenzen—termed a relativity authority by none other than Albert Einstein—and Griffith C. Evans, the mathematical guru whose name now graces Evans Hall.

From the beginning, my dad was obsessed with unified field theory and the riddle of gravitation. I was eight years old when he told me rockets would never truly open the space frontier, and it would be far better if we could make things easier by controlling gravity. His ideas weren't just advanced—they were science-fiction, but solidly backed by hard science. When you try to imagine someone who might have cracked the riddle of gravitation and invented gravity control, my father would be a most likely candidate: a mathematical physicist thoroughly grounded in relativity and particle physics, an iconoclast, a dreamer, and a rigorous follower of the scientific method.

In H.G. Wells' novel *The Time Machine*, the Inventor tries, unsuccessfully, to convince a group of skeptical friends that he has invented a time machine. Then he vanishes into the future, leaving behind one colleague, the Narrator, who witnessed the time machine's exit. The Narrator spends the rest of his days trying to convince the world that the inventor succeeded. He fails, by the way.

When I was ten years old, *The Time Machine* was my favorite novel. Even then I saw parallels between the Inventor and my father, most notably his use of special relativity. I could hardly have imagined that today I would find myself in exactly the same position as the Narrator, pleading with an indifferent world to repeat my father's 1994 experiments and see for themselves: gravity control is real, and it offers us a choice between doomsday and the stars.

Chapter 1

"Preposterous!"

SIR ISAAC NEWTON inaugurated the scientific study of gravitation in the 1600s. Today, the behavior of gravity and its inseparable companion force, inertia, are well-understood and mathematically predictable. Nevertheless, their *physical* origin remains a great mystery. Unlike electromagnetic fields, which also act at a distance, gravity has proven uniquely impervious to *any* attempt at manipulation ever since the dawn of history. Nor has a firm connection between gravity and the other fundamental forces of nature ever been established. To delineate such a relationship would be the Holy Grail of physics: a *unified field theory*.

If a reputable physicist had formulated a unified field theory and used it to control gravity and inertia with present technology, then surely his accomplishment would have been celebrated by all. It is simply *preposterous* that such a breakthrough would have been left sitting on the shelf in plain sight, gathering dust, for thirty-seven years and counting.

Yet *that*, this book contends, is *exactly* what happened. Far from being kept secret, the theory and applied technology described in this book have been advertised in public documents ever since 1981. The how and why of this story is the subject of *Book I*. As you will see, no conspiracy was needed. Institutional inertia was *more* than enough to stall gravity control on the runway. *Book I* portrays the revolution that awaits if gravity control escapes the bonds imposed by the current paradigm, including the conquest of space, a revamping of terrestrial transportation, and an economic boom the like of which has never been seen before. Even the fossil fuel industry stands to benefit. The tragedy is that gravity control has been on hold for so long. If it had been tested in 1981, it is quite likely that global warming would never have become an issue. When I was a child, my dad told me about the trouble Galileo had with the Church, and I laughed, thinking that nothing like that could happen today. Five decades of experience with today's priesthood has taught me how naive I was.

In *Book II*, you will discover exactly how the technology works. Why would I give away a trillion-dollar technology for nothing? Because the inventor gave it all away himself in a 1981 paper and several other publications afterward, rendering his invention "prior art," that is, not patentable, while the benefits to society would be enormous. How can you make money on gravity control? That question is easily answered, but before we go on, please read the DISCLAIMER that follows.

Chapter 2

DISCLAIMER

THIS BOOK IS INTENDED TO BE READ AS A WORK OF SPECULATIVE SCIENCE: THE PREMISE AND THE CONCLUSIONS IT CONTAINS MAY OR MAY *NOT* BE TRUE. The Editor (David Alzofon) encourages the reader to **remain skeptical**.

The Editor readily acknowledges a bias in favor of Dr. Frederick Alzofon's theory of gravitation and the technology of gravity control derived from it. Dr. Alzofon was the Editor's father, which makes the Editor's judgment suspect, or well-informed, or both. The reader must be cautioned that *proof* of the theory and applied technology have yet to be established by mainstream science. Indeed, the primary goal of this book is to inspire readers who possess the expertise and wherewithal to *duplicate* the experiment the Editor's father conducted in 1994. If successful, a repeat of the 1994 experiment would remove all doubt about the efficacy of the technology. It would also confirm the validity of the physical theory behind it and touch off a second Industrial Revolution. Dr. Alzofon considered the experiment a success, and, based on a lifetime of experience, the Editor accepts this judgment, while at the same time acknowledging that he (the Editor) is not qualified to render an expert opinion.

The positive effects of a successful repeat of the 1994 experiment can hardly be overstated, which explains the Editor's zealous advocacy of the theory, technology, and proof presented here. However, no amount of confidence or rhetorical fireworks constitutes scientific proof. Therefore, a reader interested in duplicating the experiment must heed the following warning and proceed with caution:

THE EDITOR, AUTHOR, AND PUBLISHER MAKE NO GUARANTEE OF SUCCESS FOR THE TECHNOLOGY DESCRIBED HEREIN, AND DO NOT ASSUME AND HEREBY DISCLAIM ANY LIABILITY TO ANY PARTY FOR ANY COSTS, LOSS, DAMAGE, OR DISRUPTION CAUSED BY ERRORS OR OMISSIONS IN THE TEXT OR BY FAILURE OF THE TECHNOLOGY TO PERFORM AS DESCRIBED. **THE READER(S) ASSUMES ALL RISK FOR COSTS, LOSS, LIABILITY, DAMAGE, OR DISRUPTION OF ANY KIND,** SHOULD THE READER CHOOSE TO ACT ON INFORMATION PROVIDED HEREIN OR IN ANY WORKS DERIVED FROM THIS TEXT IN ANY MEDIA WHATSOEVER.

An additional reason for this DISCLAIMER is that the author has no control over the quality or accuracy of the attempts of others to duplicate the experimental apparatus, which has a rather high bar of engineering precision. Because of the high engineering standard, **even minute and seemingly trivial deviations from the specifications provided herein are likely to produce failure.**

Anyone interested in replicating the 1994 experiment is *strongly* advised to seek expert assistance before investing in such an effort. We also advise seeking an independent evaluation of the science presented in *Book II* before proceeding. However, let the reader be cautioned that not all experts are equally objective: In some quarters of the scientific community, strong, even irrational prejudices are likely to be encountered. (For clarification, see *Chapter 38*, p. 241.)

THE TECHNOLOGY DESCRIBED HEREIN HAS BEEN ON THE PUBLIC RECORD SINCE 1981, when it was published in a 33-page paper and delivered in a lecture as part of the proceedings of the AIAA's 17th Joint Propulsion Conference. The original paper has long been available as a download through the AIAA website. The theory and technology have been described several times in refereed journals and magazines since 1981 (see p. 259 for list), and the 1994 experiment has been referenced in several publications. Nothing in this book can remotely be considered classified.

THE THEORY AND TECHNOLOGY DESCRIBED HEREIN WAS RESEARCHED AND DEVELOPED SOLELY BY DR. FREDERICK ALZOFON, A PRIVATE INDIVIDUAL. At his death, custody, copyright, and ownership of the material in this book passed to his heirs. None of Dr. Alzofon's work product is now, nor has it ever been the property of any corporation, government agency, or private individual other than Dr. Alzofon. All of Dr. Alzofon's concepts and discoveries in theoretical physics and applied technology are original and unique: none were borrowed or derived from the work of any other individual or group. The record supporting this, which is elaborated in the *History* section of this book and elsewhere, has a continuous record dating back to the 1940s.

PATENTS: Dr. Alzofon filed a patent application for the technology described in this book in 1980. The patent application was turned down and went into suspension afterward. **PUBLICATION OF THIS BOOK DOES NOT CONSTITUTE A SURRENDER OF DR. ALZOFON'S PRIOR PATENT CLAIM,** especially since we believe ("we" being Dr. Alzofon up until the time of his death, his patent attorney, other experts, and the Editor of this book) that the application was treated prejudicially and denied for insufficient cause. (See p. 269 for more information.)

THE EDITOR, AUTHOR, AND PUBLISHER DISCLAIM ALL RESPONSIBILITY FOR ERRORS, OMISSIONS, OR CONTRARY INTERPRETATIONS OF THE MATERIAL CONTAINED HEREIN OR IN MATERIAL DERIVED FROM THE CONTENT OF THIS BOOK. The Editor has made diligent efforts to check all facts and accurately transcribe Dr. Alzofon's work from the original sources. Nevertheless it is possible, especially in a book of this nature, for inadvertent errors to appear. Errors may also have appeared in the original source material and gone undetected by the Editor. It is the reader's responsibility to verify any areas of concern with third-party experts. The author strongly recommends consulting a PhD in electrical engineering, preferably in microwave technology or electron paramagnetic resonance (EPR) or dynamic nuclear orientation (DNO), in any serious effort to replicate the experiment.

The Editor will make an effort to answer sincere technical inquiries addressed to info@klafraknar.com, **IF and ONLY IF** the writer sends verifiable contact information, including name, address, telephone number, title, company or educational institution, and purpose in writing. The Editor will not respond to correspondence that fails to provide this information. *No exceptions.*

NOTE TO ACADEMICS: PROCEED AT YOUR OWN RISK. Anyone advocating investigation of Dr. Alzofon's unified field theory in an academic setting should be forewarned that opposition, particularly among experts in general relativity, will be intense. Negative career consequences are possible, even likely, at least until positive experimental results are published in peer-reviewed journals. Proceed with caution. This warning is unnecessary with respect to Dr. Alzofon's theoretical work in optics, heat conduction, and diffraction analysis by Sommerfeld's Method, all of which has been found as praiseworthy as it is innovative, and none of which has provoked anything like the unique response to his papers on gravity and unified field theory.

Chapter 3

Why a New Edition?

IN 2017, the first edition of this book was published under the title *How to Build a Flying Saucer (And Save the Planet)*. At the time, my frustration with institutional science and industry was as high as it has ever been. After fifty years, all the normal channels of communication had been exhausted, and I thought that the title—with its echoes of 1950s sci-fi pulp fiction—might further the cause by stirring up some controversy and triggering a media snowball. Eventually *someone* would listen, the 1994 experiment would be redone, and gravity control would go mainstream.

At the end of a year, I decided to pull the book off the market to make a few changes, beginning with the title. The new title, *Gravity Control with Present Technology*, is patterned after my father's 1981 paper, "Anti-Gravity with Present Technology." It reflects a commitment to straightforward presentation of the physics, which is how my father would have preferred it. I have streamlined my commentary on the implications of gravity control, which makes up the bulk of *Book I*.

The new title was actually my first choice when I began writing the book. But, while an improvement, it is *not* a solution. As my father discovered after publishing "Anti-Gravity with Present Technology" in 1981, mere *mention* of the word "antigravity" is fatal to serious discussions with the gatekeepers of science and industry. "Antigravity" is a trigger word for them, and they immediately suspect anyone who dares to utter such a dubious term in respectable circles. As my dad used to say, "They cock an eyebrow at you and smile, and that's the end of it."

As soon as he discovered the prejudice against "antigravity," he switched to "gravity control." Unfortunately, the new term was but a slight improvement over the old. In 2000, when I was working in Silicon Valley, I had a phone conversation with Steve Jobs about my dad's work. "What you're talking about is *antigravity*, isn't it?" he said, and I felt the oxygen go out of the air. A little while later, he asked, "If there was something to this, then Stanford would have done something about it, wouldn't they?" I had expected to be talking to him about a visionary technology. Instead, we were going down a familiar rabbit hole that would consume all the time we had. I was prepared to conduct the tour, but I could already sense his impatience. The phone call came to a polite end soon after.

The answers to Steve's questions will be found in this book, but there is no 90-second version. The technology comes with a learning curve, and that's all there is to it. My conversation with Steve was a high point of hope, but it ended like all the others between 1981 and 2017. My only goal in talking to him was to get him to talk to my dad, as anyone who talked with my father quickly realized he *did* know what he was talking about. At that, I failed, but such a failure was not unusual. Only a handful of people got that far. As for my father, he had no use for salesmanship. He spent the last 18 years of his life writing and publishing papers and books, and gravity control had little to do with any of it. Such is the "curse" of *antigravity*, or *gravity control*, or whatever you care to call it: You can't use these taboo words in talking to people who make the decisions. But what alternative is there?

The word "antigravity" is just one barrier (imagine that—a single word deciding the fate of humanity). Two others are far worse. The first is "flying saucer." The second is "UFO." Two of my father's articles about gravity control were published in the *MUFON Journal*, which a Google search would disclose in a quarter of a second, so it was inevitable that "UFOs" would come up in connection with his technology. That was another reason for the title of the first edition: If it was going to come up no matter what I did, I wanted to tackle it head on. That wasn't much of a solution, as it turned out.

Why are flying saucers an inevitable part of this discussion? Simply because a *saucer*, or *any* vehicle with axial symmetry, happens to be the ideal embodiment of gravity-control propulsion technology. Just think of a flying saucer as two dish-shaped radar antennas clapped together face-to-face. As will be explained later on, if you add a steady magnetic field and pulsed microwaves transmitted from a core generator through the top center, you get a gravity-free, interplanetary space shuttle. The saucer just happens to be the most efficient way to deliver the effect.

Flying saucers and UFOs go together like ham and eggs, but if you disregard the "aliens" and concentrate on flight characteristcs, you find that UFOs offer tangential evidence for the propulsion technology invented by my father, and vice versa: His technology argues for the existence of UFOs. This was the justification for a lengthy middle section in the first edition called "The Top-Ten UFO Riddles." I was never comfortable with it, but again, I saw no alternative, as any attempt to conceal the connection would have seemed shifty.

Immediately after publishing the book, I began to question my judgment. Like clove or curry, UFOs contaminated everything. *What was the book about? Physics or UFOs?* Another reason emerged in a conversation with MUFON Executive Director Jan Harzan, who informed me that while I had been locked away in my study writing, the powers that be had decreed the term "UFO" was *out* and "UAP" (Unidentified Aerial Phenomena) was *in*. When I heard Hillary Clinton use the term "UAP" on a late-night talk show, my heart sank. I realized that "The Top-Ten UFO Riddles," while well-motivated, should be torn out and turned into a separate book, a task I put off for a year.

As soon as work on *The Top-Ten UFO Riddles—Solutions from Science* was complete, I pulled *How to Build a Flying Saucer (And Save the Planet)* off the market and began an overhaul. The result is a leaner, meaner, more focused book. UFOs no longer appear, except here and in *Chapter 13*. This places the technology in the foreground, where it belongs. If you're interested in UFOs, by all means read the companion volume. Incidentally, the covers of both books are now almost the same.

Here's a description of *Book I* and *Book II*:

Book I is exclusively my writing. It explores the implications of gravity control for society, the economy, and the environment. There is also a brief history of the invention. Readers need to understand why gravity control is still awaiting its Kitty Hawk.

Book II is mostly my father's writing. Exceptions include *Chapter 17 – Gravity Made Simple*, a quick-and-dirty introduction to the technology, and *Chapter 30 – Theory Matters!*, a comparison between my father's work and others who claim success in this area. This chapter gives the criteria for a scientific theory and explains how these criteria can help us choose a candidate.

Chapter 38: A Word About Expert Opinion, discusses the academic response to my father's unified field theory, which can hardly be described as "scientific." This chapter has been included reluctantly, but it is essential, since anyone interested in redoing the 1994 experiment is likely to seek an "expert opinion" of my father's work, and they need to know what they are likely to hear and why. While this chapter will be read as critical of academia, it is actually rather restrained. Let me hasten to add that only in *one* area of physics were problems of this kind encountered. Everything *else* my father wrote, including for the *Physical Review*, was accepted without question. Unfortunately, the troubled area is where the arbiters of truth live, and they determine the limits of gravitation research. My father is not the only scientist to experience problems of this kind, but he is probably the most qualified physicist ever to find the door barred to publication. Evidence of a successful experiment made no difference. With this group, pedigree comes first, quality of work a distant second (see Thomas Kuhn for more).

The experimental report is included in this edition, but I have eliminated most of the illustrations. The photos were of poor quality to begin with, and the printing process made them look worse. They will not be missed, as anyone who redoes the 1994 experiment will have to redesign it using digital equipment (the original was analog). Modern computers will allow them to model the effects before they mold a single part, and that's how it should be. For one thing, the sample used in 1994 weighed slightly more than a gram. The next generation experiment should be with a metallic object of significant size and weight. As one Hollywood producer said to me: "Can you make something float?" This is the same question the Air Force colonel asked my dad in 1960. Clearly, if one wants to attract investors, "making something float" will be a desirable, if not a *key* element of a demonstration.

Finally, I have pared the discussion of the theory and the rationale for the experiment down to one chapter, the 1989 proposal. This chapter will be above the heads of some readers, so I have included numerous appendices on the physics behind the invention, ranging from "easy" to "difficult." My father's last paper on his unified field theory and dark matter will be above the heads of almost all readers, but it has been included to show his depth of learning and originality of thought.

Chapter 4

To Boldly Go...

IN SHORT ORDER, **gravity control** will open the space frontier and transform our familiar cityscapes of today into something out of *Star Trek*. For those seeking a new frontier and a revitalized economy, the news could not be better. Before we begin to explore the implications of gravity control, however, let me assure the skeptics in the crowd that we will respond to a criticism that is sure to arise, namely that gravitational fields and electromagnetic fields interpenetrate freely with no effect on each other whatsoever, yet here we are, claiming to modulate gravity via electromagnetic fields.

The explanation will be found in *Book II*, but for the record, we *agree*, and we recognize an obligation to answer in terms of known physics, not speculative doubletalk. But for now let's put the science on hold and pretend we've solved the problem of gravity control. Call it "sci-fi" if you like, but don't worry, we will provide more than enough science to support our claims later on.

Now that we're in the imaginary world of a "thought experiment," I think we can agree that the ability to control gravity and inertia allows us to do some astounding things. We can exponentially improve the effectiveness of rockets, for example, because we can temporarily make payloads—whole vehicles, in fact—lighter than a feather. And if we can do that, we can launch a spaceship with "tons" of cargo into orbit with a can of hairspray rather than the customary thirty-story rocket. NASA's current estimate for lifting a Big Mac into space is around $3,000. With gravity control, that cost will plummet to a few cents.[1]

Gravity control lifts the cap on payload *size*, too, which makes it easy to build factories, colonies, and space-mining operations on the doorstep to the stars. And, since non-aerodynamic vehicles like ours are immune to the vicissitudes of weather, our launch window has expanded to any time, day or night,

---

[1] Physicists note: We are well aware of the laws of thermodynamics. In our model we temporarily dampen the energy in earth's gravitational field. This requires an energy investment. When we shut off our device, a normal state of affairs resumes and the object recovers its weight. If the object has gained or lost altitude, hence potential energy, there is no violation of the law of conservation of energy, since this applies to closed systems, and we are operating in an open system. Gravity control transforms energy, rather than creating or destroying it. More exactly, the process is an order-disorder transformation, not unlike adiabatic demagnetization of paramagnetic salts, which has proven to be quite useful in cryogenics. The process we use is similar, but operates on a subatomic scale with the gravitational field. The exact energy requirements have been calculated, and are surprisingly low (see pp. 120 – 21).

from virtually any location. Runways and launch platforms will be unnecessary. Any level patch of ground, even a clearing in a forest, will do. Noise level: close to zero. Exploration of the planets will get much, much easier, and space tourism will become a burgeoning industry overnight.

Back on earth, we can clap a low-powered gravity-control unit on an automobile and spectacularly increase its mileage. But why limit our thinking to wheels when we can build gravity-control vehicles propelled by small electric fans or jets and fly to work in the morning at any altitude up to 14,000 feet, at almost any speed. If we install oxygen equipment and pressurize the cabin, the ceiling lifts to 50,000 feet or more. Not everyone will want to take to the air immediately, so the major auto manufacturers will be working overtime to create gravity-control vehicles whose flight characteristics fall within the comfort zone of those accustomed to four wheels planted firmly on the ground. These new vehicles will resemble the familiar "landspeeders" of the *Star Wars* saga. The effect on the economy will be staggering—a complete renovation of world infrastructure that will take a hundred years.

High-performance vehicles such as we're contemplating will require sophisticated onboard electronics, which is a good thing for commuters, as well as the computer industry. For example, you might live on top of a mountain in the Rockies and commute to Los Angeles every morning on autopilot in your personal saucer (see *Chapter 5*, p. 19). If your saucer is in the shop, no problem—just call "Uber Saucer" and one will pick you up shortly. We estimate that the cost of a personal saucer will rapidly come out of the stratosphere into the range of a light plane. Saucers without pressurized cabins will soon be in the same range as an SUV, and one day, even lower. The engineering and materials for a small saucer is quite comparable to a modern automobile, and there are fewer moving parts.

The maneuverability of gravity-control vehicles will be unprecedented. With current technology, inertia compels airplanes to make wide, banking turns. When jet fighters ignite their afterburners, inertial resistance—the dreaded g-force—crushes pilots in their seats and contorts their faces like rubber masks. Too much g-force and a pilot will pass out or die. But with inertial forces canceled, gravity-control pilots will make right-angle turns at supersonic speeds and *not feel a thing*, not even a little whiplash. The aerospace industry should welcome these developments, because it means unprecedented funding for R&D, a flood that will go on for decades.

Of course, we can't leave split-second control of airborne gravity-control vehicles to fallible human beings. Instead, we'll put the details of flight in the hands of onboard computers linked to a central traffic-control network. With flight paths subject to security review and failsafe systems onboard, gravity-control vehicles will be less likely to be hijacked (or hacked) than modern commercial aircraft. The security issues are complex, but solvable. Suffice it to say, parents will be able to trust their teenagers to take the saucer for a spin on a Friday night, because the kids can determine the destination, but they won't have access to the gas pedal or the steering wheel.

Nearly every engineering equation ever devised has to take gravity and inertia into account, which is why gravity control will have such a liberating, transformational effect on society. Today, gravity and inertia are considered more reliable than death and taxes. We take their influence for granted, and have done so ever since dawn of history. *Nothing* escapes the influence of gravity—not the trees, or birds, or blades of grass, the planets, the sun, the stars, or galaxies. Fundamental concepts such as "up" and "down," "heavy" and "light," are all dependent on an intuitive grasp of gravity and inertia. The way we walk or run or kick a ball, too. That's why the chasm that separates the modern world from the horse-and-buggy era is *nothing* compared to what will come after gravity control. A portrait of the future can be found in *Chapter 5*, but the point here is simply that once the technology begins to spread, it will quickly pervade every corner of society, from where we live and how we get around to the way we look at the stars above and the earth below. *Everything will change.*

And everywhere that gravity control is used, it will reduce our reliance on fossil fuel. Consider it a fringe benefit that will save the planet. If you can commute to work at fantastic speeds in a gravity-control vehicle propelled by a few small electric fans, how many gallons of fuel will you consume each week? You will need to charge the battery or exchange it for a charged unit at your destination, but the charging can be done with solar, geothermal, or hydroelectric power, and the mileage you get on one charge will be far greater than a conventional four-wheeled vehicle. Hybrid vehicles that use airflow to drive a turbine may relieve power requirements (see p. 26). You'll spend much less time traveling than before, because your gravity-control vehicle will get you to your destination at hundreds, if not thousands of miles per hour.

You will never have to hit the brakes for a traffic jam, either. Gravity control will tilt today's five-lane highways into the sky, with slower lanes nearer to the ground and faster lanes higher up, and no stoplights anywhere, anytime. The low-altitude route into town might take longer, but the scenery will be spectacular, as you flit over mountaintops and through the canyons like a hummingbird. It's inconceivable to us now, but in a hundred years, the asphalt and concrete freeways we take for granted today will likely revert to horse trails, public parks, and forests.

Natural power sources, such as wind, solar, hydroelectric and geothermal, have already reduced consumption of fossil fuel and will surely reduce it more in the future, but none of these alternatives have directly affected the use of fossil fuel in transportation. Gravity control will have a profound effect on carbon emissions by increasing miles per gallon while decreasing transit time and gradually edging out fossil fuel. If the batteries or capacitors that energize gravity-control vehicles are charged by natural power sources, such as solar and geothermal, then global fossil fuel consumption will soon diminish to safe levels. The decline in aggregate demand will be the same as if we had installed a mileage-saving electric propulsion system on the entire planet.

Before Big Oil puts a rattlesnake in my mailbox, they should read *Chapter 6*, because gravity control will make them richer than ever while taking care of their most vexing problems.

Skeptics will no doubt question how much energy it takes to induce the gravity-control effect. If the energy cost was *high*, our scenario would lose its luster. However, my father made some calculations in the lead-up to the 1994 experiment, and the figure was surprisingly low. This is because the technology resembles a pump: A certain amount of nuclear orientation is retained between microwave pulses, and since the pulses are cycled extremely rapidly, the effect builds up quickly and diffuses throughout the hull of the craft. On this basis, he predicted that the weight loss would be tentative at first and would suddenly cascade downward to the gravitational equivalent of absolute-zero degrees Kelvin. He also predicted that the pumping action means a low-power input can be leveraged into a great weight loss. The cumulative effect he predicted *was* observed in the 1994 experiment. (For precise calculations, consult p 120, and the 1981 paper, pp. 4 and 5.)

A complete solution to the problem of energy production is not within the scope of this book. Our purpose is only to spread a technology that will cut consumption of fossil fuel to safe levels within the near future, not eliminate it entirely. That's not the whole story, however. In the 1989 grant proposal (*Chapters* 18 – 19), my father pointed out that his unified field theory has a direct bearing on future research:

Energy Production

Since energy cannot be created or destroyed, the "production of energy" is a term which must refer to the conversion of one kind of energy into another, and useful, form of energy. Often the conversion takes the form of transforming a potential energy into a useful kinetic energy. For example, potential gravitational energy is transformed into electrical energy by use of falling water, or a gas under pressure may expand to cause motion of a piston.

Approximately characterized, the history of energy generation has progressed from an original dependence on natural sources such as muscle and wind power, etc., to use of exothermic chemical reactions, then to production of electrical power, and finally to production of subatomic reactions (nuclear reactors, fusion, etc.). In general physical terms, this progression has been toward reliance on physical processes involving energy stored in smaller and smaller entities and greater power yields. But in each case, finding a useful energy transformation depends on knowledge of particular details of a specific process which makes the desired energy conversion possible. Only experiment can disclose these details; however, experiments cannot proceed without an adequate theory to guide the way. This theory must tell how to convert the random (or probabilistic) motion of atomic and subatomic entities to macroscopic, ordered motion for extraction of useful work.

At this time, further progress in employment of subatomic processes for extraction of power is hampered by lack of understanding of subatomic processes. Although much is known from experiment, the theory of such processes is admitted to be inadequate. This lack of complete understanding is illustrated by the numerous infinities that result from modern theory, e.g. the concept of an infinite amount of energy stored in a "vacuum."

It has been hoped that discovery of a unified field theory will resolve the difficulties inherent in present models of reality; it is the writer's contention that his unified field theory accomplishes this role, and will show how the desired energy conversions can occur. [See pp. 109 – 111 for context.]

He was probably entertaining notions of how subatomic processes might be utilized at the same time he wrote these words. During one of his rare visits to the Bay Area in the mid-1980s, we were driving north along a scenic stretch of Highway 280 near Crystal Springs reservoir, when he looked out the window and started to describe an idea he'd had for a new kind of power source, a kind of "nuclear transistor" that would boost the random electrodynamic activity of subatomic particles into usable energy, much the same way a transistor boosts a signal in a circuit. Over the years I frequently asked him whether he'd made any progress on the idea. In the 2000s he told me that he'd given up on it because it was impossible, but he never gave me a reason why. The notion of "free energy" was unpalatable to him, so perhaps it was the law of conservation of energy that ultimately convinced him that the "nuclear transistor" had no future. But maybe this anecdote will stimulate the imagination of readers who take the trouble to acquire a thorough grasp of the UFT (Refs. 1a, 2, 4, 7, 9, p. 259).

While gravity control does not offer a direct solution to the problem of energy production, the advent of cheap and easy space travel, plus the ability to lift vastly increased payloads into space with far greater frequency, will open the door to new power sources here on Earth. The outlook for solar power has improved in recent years, but solar radiation encounters many impedances before it reaches the surface of the planet, including the angle of incidence, the air mass, dispersion from atmospheric gases, and the simple issues of day and night, cloudy or clear. In effect, solar cells at sea level are drinking from a solar soda straw. In space, no such impedance exists. There is as much solar power as one can gather, twenty-four hours a day. With gravity control, it would be possible to place enormous solar collectors at Lagrange points (orbital sweet spots) and collect enough solar power to run whole cities. The question is how to relay it back to Earth? Obviously cables will not do.

Conversion of solar energy to microwaves might be one way.

On November 7, 2013, gizmag.com reported a breakthrough in harvesting electricity from ambient microwave energy *already* available in the environment. This suggests that solar collectors could beam energy back to Earth in the form of microwaves, and the energy could be converted back into electricity at suitable collector sites here on the ground in remote places such as the Mojave Desert.

In March, 2016, the Space Solar Power D3 (diplomacy, defense, development) team, which included Dr. Paul Jaffe of the Naval Research laboratory, won an award from the National Space Society with their proposal for a system of space-based solar energy collectors (a seven-minute video is available at the National Space Society website, nss.org). The winning team included members of the Air Force's Air University, the Naval Research Lab, Northrop Grumman, NASA, the Joint Staff Logistics and Energy Division, the Defense Advanced Research Projects Administration, the Army, and the Space Development Steering Committee—all respected organizations with impeccable scientific credentials.

My lack of expertise prevents me from speculating further, but the point is that gravity control would facilitate the development of the D3 plan, while it accelerates the pace of space exploration and opens new possibilities we can't even dream of within the current framework of space technology.

Gravity control will help in other ways, such as allowing millions of people to view the Earth from space, which will change how we perceive humanity's place in the cosmos. See the next chapter (p. 32) for more.

The technology will also allow us to permanently remove toxic waste from the environment and protect the planet from rogue asteroids or comets. Cheap and easy space travel will lessen sources of conflict on earth by opening up a new frontier, reducing population pressure, stimulating the global economy, and replacing the culture of scarcity with a culture of abundance. Pessimism will fade, optimism will grow. All of these will rapidly transform society in a positive manner. (Also see *Chapter 5*, "Beyond the Scenario," p. 30).

And again, Big Oil will not be left out. In fact, the fossil fuel industry is likely to be an early adopter of gravity control. See *Chapter 6*, p. 35.

Chapter 5

A Trillion-Dollar Technology

"MULTI-TRILLION" would be more like it. My dad estimated the value of gravity control as equal to the combined output of all the industrialized nations on earth within ten years of its debut. Nevertheless, in Palo Alto one day, a venture capitalist listened to my pitch, including points made in *Chapter 4*, gave me a skeptical look, and asked, "Just how do you propose to make money on this?"

After I picked my jaw up off the floor, I told him, but he remained skeptical. I soon found that this question was common among investors, so let me put it to the reader: Can you see any way that someone might make any money from gravity control? One condition: You can't use the patent on gravity control itself, because after 1981, it was prior art. Can you write down three examples?

Done? All right.

Some of you probably had a hard time listing anything at all, even after reading *Chapter 4*. If so, you're in good company. Mark Twain thought that Alexander Graham Bell would never make anything with that gizmo called the "telephone," and Twain was the author of *A Connecticut Yankee in King Arthur's Court*, a book about a time-traveling engineer marooned in medieval England. Twain was a genius, but he got it wrong about technology, not once, but many times. Think for a moment how many different ways telephones make money now. For one thing, there are millions of patents on different kinds of phones and the parts inside those phones, as well as the infrastructure of cellular phones and land lines. Then there's the money made by the telecom companies. And on and on.

Gravity control will evolve along similar lines:

• If Mattel can make a fortune on *toy* spaceships and automobiles, what will someone make by mass-manufacturing *real* spaceships and air-cars that cost the same as an SUV?

• What could be made by transforming the auto industry everywhere on earth?

• What business opportunities will arise in public and private transportation?

• What could be made on a space-based factory the size of Manhattan, or a space-based resort, or a tourist industry based on interplanetary travel?

- What about space mining? I hear there are oceans of water on Europa, Ceres, Pluto, and who knows where else. We need water here, and new colonies on the Moon and Mars will need it too.

- And what about government and military contracts?

- What about *Disneyland*, for that matter? What kind of amusement park rides could be designed? You mean people would actually pay *money* for such rides? Nah, of course not.

- What will the advent of all these new industries do to the level of economic activity all over the globe, not to mention off-world colonies and mega-space-stations?

All the traditional avenues to wealth generation are open, from products and services to patents, royalties, and licensing agreements. Gravity control will also generate secondary industries all over the world, and for the first time, in space. These are the very things that power the world economy today, but the future will magnify them to an unimaginable scale.

I don't know for sure, but I have a sneaking suspicion there might be a dollar or two to be made in there somewhere. *Maybe*, if I squint my eyes real hard and peer into the future.

The money question is revealing, because it suggests how *incomprehensible* a radically new technology can be to people whose minds are immersed in the "matrix" of things as they are today.

Seeing outside the matrix isn't easy, as I learned first-hand in Silicon Valley. In the mid-1970s, I took a tour of the Xerox PARC laboratory on Coyote Hill Road in Palo Alto in the company of an old high school friend, Marc LeBrun,[2] who showed me a prototype bit-mapped screen and a mouse. Both were entirely new, and the technology was primitive by today's standards. It is embarrassing to admit, but on the day of my visit, I saw only what appeared to be an electronic Etch-a-Sketch. "What would anyone *do* with something like that?" I wondered.

Ironically, the same bit-mapped screen and mouse were to provide me with employment in 1985, when I was hired as a technical writer by Jef Raskin, the genius behind the Macintosh project at Apple. Jef, who was doubtlessly a genuine visionary, launched the Mac project after seeing exactly the same device I had seen at Xerox PARC ten years earlier.

Steve Jobs didn't see the mass-market potential hiding out at Xerox PARC, either. When Jef explained the concept of the Macintosh to him—a versatile, low-priced computer with a bit-mapped screen and a mouse, a kind of "information appliance"—Jobs said (according to Jef), "That's a *stupid* idea." The pricey Lisa was the Apple of Steve's eye at the time. A few years later, he realized his mistake, but the point is that even someone as smart as Steve could miss the value we now take for granted.

---

[2] Tip of the hat, Marc.

Do you think *you* would have seen the value of the bit-mapped screen and mouse? Of course! It's *obvious*. Or is it? There's an anecdote about Columbus and the egg, but I'll leave it to you to look it up if you don't know it. The anecdote above shows that even visionary technophiles can get it wrong. If my experience in Silicon Valley from 1984 to 2007 taught me anything, it is that insanely great ideas don't often look that way until they are viewed in the rear view mirror, and the ideas that looked hot at first blush are often miserable failures.

Both Steve and Jef failed to see the future of gravity control. That is the reason for this chapter. It's insurance. I happen to think that gravity control will transform the world more than any invention since the wheel, but you may be scratching your head and wondering what the heck I'm talking about. By dramatizing a day in the life of a future sales executive, I hope to bring the vision to life.

A DAY IN THE LIFE OF A YOUNG, UPWARDLY MOBILE PROFESSIONAL, 2057 AD

The following scenario evolved slowly, thanks to many attempts to help Silicon Valley investors understand the implications of gravity control during the years 2000 – 2007. I was able to confirm the feasibility of most of the devices described here in conversations with my father.

"($$)" indicates products, services, or commercial activity generated by gravity-control technology.

Orientation: It has been four decades now since the technology of gravity control went mainstream after a high school student who wanted to do something interesting for a science fair followed the instructions in this book and rebuilt the 1994 experimental apparatus in his garage with the help of his father, an electrical engineer. Today that kid, now middle-aged, is one of the richest people on or off the planet. But our story doesn't concern him. It concerns someone far downstream from the early days of gravity control, someone perhaps like you.

You are a latter-day "space yuppie," a young sales executive at Interplanetary Designworks, Inc. (IDI), a company that builds a wide variety of space stations serving the needs of industry and entertainment ($$). You began working at the company fresh out of college, and seven years into your career ($$), you are just a trifle more prosperous than your yuppie counterparts in 2017.

As for the Editor, he is still around, 107 years old and sitting in a wheelchair on the lawn of an old age home in Pasadena with an IV bottle dangling over his head and a twenty-something nurse nearby, reading a magazine, but don't bother visiting—he's long since forgotten writing this book. Even the air-cars whizzing by over his head on their way to Los Angeles fail to remind him of it.

6:00 AM: The alarm goes off next to the bedstead on the second floor of your humble, ten-room log cabin deep in the Rocky Mountain wilderness ($$). Music for Zen meditation fills the air. You never had much use for buzzers. And the shakuhachi flute and meditation bells set the tone for today's business.

You roll out of bed and stagger over to the picture window, rubbing your eyes. "Curtains open," you say. The curtains roll back, revealing a panoramic view of distant jagged peaks cloaked in snow, and a pristine alpine lake surrounded by virgin forest at the base of the steep mountainside below your cabin. Thunderheads gather over the peaks on the northern side of the lake ten miles in the distance. You never tire of the ever-changing view. Right now, the billowing anvils of white and charcoal gray seem to be seething higher while you watch, as if in time-lapse photography. But it could spell trouble. You don't want this to be a work-at-home day.

"Weather at seven?" you inquire in the empty room. "Sixty-two percent chance of rain with lightning," says the home management system, which has a link to the national weather service satellite. Maybe if you hustle, you can outrun the storm. Today's a big day—you don't want to be late.

While taking a shower, you reflect on your good luck. What would you be doing now if you hadn't gotten in on the ground floor with a trillion-dollar company? Probably selling real estate in Anaheim instead of selling custom-built space stations to seventy-two different nations ($$).

The cabin was a sweet deal. Your last sales commission allowed you to build it deep inside Rocky Mountain National Park, elevation just under 10,000 feet ($$). Great bachelor pad. Peace and quiet. Fly-fishing when you feel like it (if only you had the time). All you had to do was work out a remote-dwelling lease with the National Park Service. You built a fire lookout on your property, made a hefty donation to the park service ($$), and they leased the land to you for ninety-nine years, as long as you maintained the lookout.

A lot of deals like that were struck at the beginning of the remote-dwelling craze: national parks, remote islands in the Pacific and Indian Oceans, the valleys and summits of the Andes—places where no-one ever lived before became suitable for commuters like you ($$). Building the cabin would have been impossible without gravity control, too ($$). Sitting high up on the side of a mountain? No roads? No electricity? No plumbing? But you had all of that put in ($$). It wasn't cheap, but in the old days the project simply wouldn't have happened. With gravity control it was easy to move the materials up the mountain to the site ($$) and get the builders in and out ($$). You modeled it after the spectacular Beaver Meadows Visitor Center at the park entrance. Frank Lloyd Wright was way ahead of his time. It was almost like he knew gravity control was coming.

But then there *was* the daily commute to Los Angeles ($$). Nine hundred-fifty miles. It could be a hassle, but it only took forty minutes on average. Oh well, better get cracking.

6:51 AM: Your GCV (gravity-control vehicle) ($$) sits on a platform in a clearing in back of the cabin, a dull bluish-gray. The disc is seventeen feet across, roughly the length of an old Ford F-150 truck. Strictly speaking, it isn't a saucer, but more of a croissant (see p. 168) There's a crescent-shaped cutout across the back and twin engine hubs on either side of the cockpit. The cockpit dome is the latest—that new kind of aluminum ($$). It's opaque now, but when you power up, it becomes transparent. Very cool. The engineers knew what they were doing. Just a quarter of a century ago they were making old-fashioned automobiles. But they were a forward-looking company; they moved fast to get beyond the "mileage enhancement appliance" stage, an add-on that reduced inertia in conventional vehicles ($$). And look at them now —manufacturing the best luxury GCV on the market ($$).

You're zipping your flight suit ($$) on over your business attire as you climb the steps to the platform, carrying your notepad in your other hand. Roiling thunderheads are cresting the mountains, casting shadows on the valley now, but there's no time to admire the view. Occasional lightning flashes and the distant rumble of thunder suggest you might be grounded momentarily, and you can't be late. The park ranger drops his binoculars and waves down at you from his post in the lookout platform. You wave back. Ranger Mike comes and goes the old-fashioned way, on horseback. But he does have a pair of drone orbs ($$) he can launch any time reconnaissance is needed.

The charging cable ($$) decouples from the belly of your GCV and retracts below the platform as you open the dome with a biometric key ($$) on your notepad. Solar cells ($$) collect energy during the day and begin charging the GCV's capacitor ($$) as soon as you touch down after dark. When you're in L.A., your GCV is plugged into a charger in the storage garage ($$) or you swap it out for a freshly charged module ($$). Your GCV rarely needs a charge, however, because you have a secret power source (described below).

Your GCV is a two-seater ($$). The salesman ($$) called it a "sport saucer." The model is the "Europa," in honor of Saturn's icy moon. A lot of water on Europa ($$). IDI is building a space station out there. You had a hand in closing the deal, which was worth $100 billion all told ($$).

You hop into the pilot seat and the dome closes. The airtight seal ($$) engages and the cabin pressurizes. The instrument panel ($$) lights up as the gravity-control unit switches on beneath your seat with a faint hum ($$). "Destination?" asks the AI pilot ($$). His voice was borrowed from a certain starship captain in an ancient TV series—"Picard," was it? The Shakespearean accent is supposed to inspire trust and confidence, and in fact it *is* rather effective. You've nicknamed him "Horatio," for Hamlet's BFF. This is your fourth GCV, and the best of the lot ($$). Horatio has followed you all the way.

"*Work*," you say. There's a pause while Horatio consults with the national air traffic system ($$).

"Three routes available, sir," he says.

"Display."

The holographic map ($$) inside the darkened dome shows the western United States with three zigzagging lines connecting your house and downtown Los Angeles: red, green, and blue. Each route displays the ETA at the landing area near the IDI building. All of them are under one hour. All calculated by FAA computers in under a minute ($$). No such thing as a traffic jam with a GCV; rarely a delay of any kind ($$). You've seen photos of the old days, when traffic jams turned freeways into parking lots. Thank god you were born after that time. How many years did everyone waste just sitting in a creeping rhumba line of cars, breathing noxious fumes and gulping blood pressure pills?

"*Blue*, Horatio."

"Ah, the scenic route, sir. Excellent choice."

The red and green routes disappear from the map as the hologram shrinks to an eight-by-ten-inch window on the surface of the instrument panel ($$). The dome fades from opaque to transparent and the mountains, the lake, and the towering thunderheads all come into view again, as if through sunglasses. A gauge on the instrument panel ($$) displays the weight of your GCV. It bounces around as the effect takes hold: 2517 pounds, 2550, 2100, 2213, 1625, 1685.... Warming up usually takes two minutes. Suddenly the numbers cascade downward—1200, 972, 540, 210, 87, 65, 21—a bell sounds.

"Weightless, sir," says Horatio. Your body feels like a dandelion, floating beneath your body harness.

"Lifting off."

The GCV begins to rise slowly, like a helium balloon. The analogy is accurate: Your vehicle displaces a volume of air greater than its current weight, which is nearly zero, so buoyancy lifts it off the ground. Only the faintest sensation of movement accompanies liftoff; when you're underway, there is none. It is as if the ground is pulling away from you, not *you* away from the *ground*. At one hundred feet, the air jets hiss ($$), steadying you. The gravity control system adds a little ballast to the vehicle to establish equilibrium. The Europa rotates slowly and steadily, a side effect of the gravity-control system (see *Top-Ten UFO Riddles*, p. 79). Your altitude above the mountainside is a hundred feet, treetop level. You can see the roof of your cabin below, but your altitude above the valley and the lake is 3,000 feet.

"Checking flight path," says Horatio. "Please stand by for authorization." All routine now, but still spectacularly thrilling.

This is an AI flight ($$), like taking a taxi in the old days. Except Horatio is a better driver. You can choose from a spectrum of autopilot interfaces on antigravity vehicles, ranging from a "full service" autopilot like Horatio to full-manual control, which is rare and restricted to low-performance vehicles, many of which can fly no higher than thirty feet ($$). Full-manual piloting requires special training, licensing and security checks ($$), and the interfaces are extremely expensive, which is why almost all manual pilots are military ($$). Even then, they use a hybrid smart-control system with built-in collision avoidance systems ($$). That's fine with you, as you enjoy the view during the first part of the ride, and you like to read the newspaper (paperless, of course, downloading now while you hover - $$) once you reach full altitude. Let Horatio worry about navigation, altitude, weather, traffic.

You are being tracked at all times ($$). In the unlikely event of an inflight emergency, Horatio will land the Europa—on any flat surface larger than one hundred feet in diameter—and a service vehicle ($$) will arrive within an hour to give you a ride and tow your GCV ($$). In case of gravity-control failure, a well-concealed parachute opens ($$). In the event of *catastrophic* failure, the cabin capsule ejects from the vehicle and parachutes to the ground ($$). Failures such as these occur far less often than auto accidents used to in the old days, and the injury rate is far lower—almost nonexistent, though the ride down can be exciting.

Fear of flying and fear of robots was one of the reasons it took a while for the public to accept vehicles like the Europa. In the beginning, the demand was for low-flying, fan-powered hovercraft that flew at "safe and comfortable" speeds, up to 120 mph, and had optional autopilot safety features ($$). But gradually confidence built, along with engineering standards. And now? Los Angeles looks like a beehive on any given day, with swarms of GCVs and space vehicles entering and exiting the heart of the city ($$), but the traffic flow is programmed like a Swiss watch and the air is as clean and clear as it was in the days before the arrival of civilization on the West coast.

You enjoy the 360-degree panorama of the valley provided by the lazily turning GCV for the few seconds it takes to obtain FAA clearance. The instrument panel shows that your external flight lights ($$) are blinking. Your craft can be seen from miles away. The Europa has a variety of stabilizers and backup stabilizers, including computer-controlled ($$) electromagnets, compressed gas jets, and fans ($$). But there's plenty of room onboard, since all of the stabilizers are small and weak, but their effects are magnified once gravity and inertia have been canceled.

Propulsion is provided by a pair of turbojets ($$) inside the flight stabilizer humps alongside the cabin. These clean-burning jets would have powered a small model airplane in the old days, but they are more than sufficient to get you to Los Angeles at Mach 3 now.

Turbojets are only one of a variety of propulsion systems available for gravity-control vehicles ($$), but it was decided long ago that only the military and exceptional civilian agencies would use advanced electromagnetic drive systems ($$), which were capable of Mach 18 or better. Electromagnetic drives are silent and require no combustible fuel, so they are easily capable of exiting earth's gravitational field into space. The vastly different requirements of space travel and terrestrial travel, plus the added expense of an electromagnetic drive, meant that civilian drivers were restricted to various kinds of compressed air, fans, and jets for normal use. In addition to the advanced technology, spaceworthy vehicles require special training ($$) and permits ($$), as well as heat and radiation shielding ($$), celestial navigation equipment ($$), and so on. However, civilian space vehicles are available, and you will be taking your clients up to the IDI showcase station in one of those today.

Your jets will propel the Europa at more than 2000 miles per hour. The first time a human being traveled at Mach 3 was a hundred-and-one years ago, when Captain Milburn Apt of the U.S. Air Force dropped his Bell X-2, a swept-wing experimental jet, from a carriage under the starboard wing of a Boeing-50 propeller-driven Superfortress at 31,800 feet over Edwards Air Force Base, California, and fired his rockets. The X-2 hit Mach 3.196, but the plane went into a violent roll at supersonic speeds. Apt ejected, but he was killed and the plane exploded. The reason the X-2 went into a roll was not a problem with aerodynamics. It was conservation of angular momentum. The phenomenon, which was unknown until the advent of supersonic aircraft, is called "inertial coupling." It occurs when the pilot tries to make a turn at supersonic speeds, and the inertial forces on the wings and fuselage get into a fatal argument.

Inertial forces are *unknown* on the Europa, so you can do something undreamt of in 1956: make sharp turns at supersonic speeds. *Very* sharp turns. The notion of a world in which gravity and inertia are no longer inevitable was so incomprehensible at first that private GCVs were programmed to make smooth, banking turns only, so as not to frighten the passengers. You, however, were born fifteen years after the revolution began and have never *known* a world where gravity and inertia were not malleable. For you, right-angle turns at supersonic speeds seem normal. All you had to do was push the "Accept" button for high performance and choose the parameters when you set up your craft.

Your commute craft is clean and green. Getting to L.A. will take less than four gallons of alcohol ($$). Once you're traveling at speed, thin scoops under the wing divert air to a turbine that powers the antigravity unit ($$). This allows you to stay airborne almost indefinitely. The air slits and the turbines add a muffled hiss to the background noise, but it's a small price to pay for the added range.

Aside from your mini-turbines, the electrical energy required to charge your vehicle comes from solar, wind, geothermal, hydroelectric, and space-relay sources ($$). It is stored in hyper-efficient onboard lithium-ion batteries and the main capacitor ($$), which powers the gravity-effect generator during takeoff. By 2017 ($$), the year that gravity control was released, Tesla Motors had already made significant breakthroughs in battery technology, and the forward-thinking company rode the boom, literally to the stars. It wasn't long before the rest of the world wanted in on the act ($$). Patents and licensing agreements made billions for the company's shareholders ($$).

Your flight time will be roughly half an hour, but with takeoff, landing and parking rituals, plus the time it takes to walk to the office from the GCV storage area, you will step through the door of the office at 8:00 A.M. Might even have time to pick up a cappuccino-to-go at Starbucks.

7:02 AM: The Europa swings around so that its leading edge faces west, the turbojets ignite, and you instantly accelerate to 400 mph. A camper or hiker standing below you would hear only a faint hiss as you passed overhead. If he blinked, he would miss you. Inside the soundproof cabin, you hear nothing, not even the air brushing past the hull. A side effect of the gravity control system is a layer of ionized gas on the surface of the GCV, which makes flight almost frictionless and enables your craft to reach supersonic speeds effortlessly, without sonic booms. Again, because of the absence of g-forces, it seems as if you are resting comfortably in the virtual world of a flight simulator, rather than accelerating to Mach 3.

"Canyon route, sir?"

"Sure, Horatio, but get me to L.A. on time."

"Of course, sir."

The Europa dives through a series of canyons on its way out of the Rockies. It's a route you love to take every morning, but today it keeps you below the roiling air currents above. It takes three minutes to reach the foothills. By then, the storm is behind you and the sun is at your back.

"Climbing to 40,000 feet," says Horatio. Your altitude is a function of the attitude of the craft, jet propulsion, and buoyancy; it has nothing to do with aerodynamic lift. Since your jet fuel is oxygen-dependent, your vehicle is not space-capable. Nor do you have the permits to exit the Earth's atmosphere. Your company GCV? That's another story.

Horatio angles upward to cruising altitude on a steep trajectory, but no g-force pushes you back in your seat: You're wrapped up in a news report on a Mars mining operation that discovered something interesting below the surface yesterday: *fossils!* The science is fascinating, but you see something more in that headline—a *business opportunity* ($$).

9:00 AM: Your visitors arrive at IDI, an international group that includes investors from Japan, the United States, Mexico, the United Kingdom, and France. You usher everyone into the conference room, where you are joined by the CEO and six staff members. Everyone is seated and a light breakfast is served. After introductions are made and smartphones capture business cards, the head of the delegation speaks.

"As you know," says Mr. Mikami, "We have been developing luxury resorts for thirty years. We specialize in remote destinations, Shangri-La oases in the Himalayas, Bora Bora, Patagonia, Alaska, the Canadian Rockies, Southeast Asia, the South Seas ($$), but nothing in space. Our new concept is for a kind of Zen retreat with panoramic views ($$) of the stars and the Earth ($$). We want to create an environment conducive to spiritual renewal and awakening. The idea is to cater to roundtrip customers, or visitors on their way to or from the Moon or Martian colonies. You have read our preliminary plan?"

"Yes, I have, and I think that IDI can make your wish list come true. But rather than spend the morning looking at a slideshow, why not go upstairs and see our latest project."

No special preparation is needed for the ride into space. You simply conduct everyone to the rooftop landing area, where the company's spherical shuttlecraft awaits.

9:30 AM. You arrive at the company's latest space station project ($$), 249 miles above the surface of the Earth. The flight itself took roughly three minutes, as the company shuttle uses electromagnetic drive and the vertical distance is negligible ($$). Once it becomes weightless and inertia-free, the pilot uses buoyancy to lift off like a hot air balloon to takeoff altitude, and after receiving clearance from the traffic controller, he switches on powerful electromagnets ($$) in the belly of the craft that rappel off the Earth's magnetic field. After a brief hovering period while you wait for clearance, the acceleration to 5,000 mph occurs almost instantaneously. Your guests, who are seated in outward-facing chairs in the observation room, would say they felt more g-force in an elevator. The Earth simply pulls away on the wraparound monitor as if the camera lens were zooming out. As Earth's gravitational field doesn't resist your ascent, there is no necessity for achieving "escape velocity."

Once you near the space station, which is only half complete, you have the pilot, who is certified for manual flight, maneuver around so that your guests can appreciate the sheer magnitude of the project. Slated to become earth's first full-fledged, permanent space colony and deep-orbital trading outpost, the framework is five miles long. Construction takes place in a relatively low, geosynchronous orbit, so that materials ($$) can easily be brought to the site, both from the Earth's surface and from space factories on the Moon and in the L4 and L5 industrial zones (stable orbital points on either side of the Moon) ($$). While docking, you pitch IDI's record of achievement. When the colony is complete, it will be moved 183,000 miles farther out to the L5 zone ($$).

"The spiritual dimension of space travel was one of the unexpected side effects of gravity control," you say. "IDI recognized this early on and began constructing stations to serve the needs of many of the world's great religions ($$). You are the first to bring us a project with a Zen theme, however. We would work closely with you to establish an environment that captures the spirit of a Zen temple, but with modern elements."

"Just how flexible can you be, when space imposes so many restrictions on architectural design?" asks the French delegate, a young woman about your own age. It is impossible not to notice that she is quite attractive. "Dr. Trouviere," she reminds you when she catches you at a loss for her name.

"I think you'll be surprised at what we've been able to accomplish in our latest project, Dr. Trouviere. Ladies and gentlemen, welcome to orbital station *Athena*, planet Earth's third international space colony, and, I think you'll agree, its best yet."

8:30 PM: You are seated in a cool leather chair in the CEO's office atop the IDI building in downtown Los Angeles. The night sky hums with firefly lights from private and public air traffic ascending and descending throughout the city in a constant hum of activity ($$).

"Impressions?" says your boss, handing you a scotch on the rocks.

"I think they were impressed with Athena."

"Who wouldn't be? But are they *buying*?"

"We only have two competitors for an operation on this scale ($$). And we have exclusive leases on those L5 spots ($$)."

"Did anyone mention piracy?"

"Well, yes, they did, and I reviewed our security measures ($$) as well as our contract with Space Protective Services ($$) and the United States Space Force presence in the area ($$)."

"Bottom line?"

"Mr. Mikami is staying in L.A. I'm meeting with him on Wednesday. Should have a decision then."

"Bring me up to date on the other projects in the pipeline."

"The proposal for a permanent film studio in space is still being reviewed ($$). We have got four bids on the proposed entertainment module for the Mars station ($$). The buyout of Space Transport and Construction Co. is proceeding ($$). And our subsidiary, SaucerDock, has sealed the deal on parking software and infrastructure for Boise, Cheyenne, and Santa Fe ($$).

"Good work. Why don't you take the weekend off?" Your boss smiles with a sardonic smirk. He knows how many weekends you've spent with nose to grindstone—the price of a booming market.

"Well, there's one more thing. Did you hear about the fossils on Mars?"

"Yeah. Big news."

"If we move quickly enough, we could establish a museum base near the site, say, as an entertainment spot for miners, colonists, and scientists, and piggyback that with expeditions from the colonies or from earth to 'see the fossils live'—in situ, that is ($$)."

"Sounds possible. Can you work up a report by next Friday?

"Done."

9:40 PM: You are cruising eastward at Mach 3, 40,000 feet over central Utah. Your ETA in New York is one hour, but Paris is your ultimate destination. Dr. Trouviere, or "Corinne" as you've come to know her, is strapped into the copilot seat on your right. She is hitching a ride back to the Continent. There's been a lull in the conversation, and you are tempted to fill it.

"You know where you want to eat breakfast?"

"Mm, there's a little place near Canal Saint Martin ($$)."

In a few seconds you find the landing site in Paris and book a reservation ($$).

Corinne smiles. "Don't worry, you are off the hook."

"I am?"

"For being late. I would say a weekend in Paris ($$) more than makes up for it."

"I thought we'd stop off on the way there ($$)."

"Sure, why not. Anywhere, so long as it's not trout fishing. I know you like it, but I am sorry. Where did you want to stop?"

"New York, and in the morning, Turks and Caicos."

"Where is Turks and—what is it?"

"*Caicos*. It's in the Caribbean. A set of coral islands. An archipelago. Most of those islands would be underwater right now, you know, if global warming hadn't been stopped."

"I don't know, all that was before my time."

"It used to be a big deal. Not too long ago, either."

"Of course, but you know what is a big deal for me?"

"I can't imagine," you say, glancing across the cabin at her. Corinne smiles impishly, a bang of auburn hair veiling her right eye.

Suddenly you both laugh. Finally, work is behind you. It's going to be a glorious weekend.

BEYOND THE SCENARIO

To get an idea of the social and economic impact of gravity-control technology, one need only begin adding the phrase "gravity control" to any engineering problem where gravity and inertia enter into the equations. This includes trains, planes, drones, automobiles, turnpikes, ski lifts, amusement park rides, cranes, forklifts, dry-dock facilities for boats, yachts, ocean-liners, tankers, and high-speed machinery of all kinds. All of these are moneymakers, and as the technology spreads, the mint will be burning up to keep pace with the economic expansion.

As the technology spreads, the world will enjoy an economic boom far greater than that caused by personal computers and broadband technology. We will see a mass mobilization of industrial resources not unlike World War II, but without the bombs or battlefields. One might argue that when fewer workers can move more freight, there will be fewer jobs, but we would argue the opposite, because projects that have been held back by high costs or engineering barriers will suddenly become feasible, as well as projects currently beyond imagining.

Another source of jobs will be the overhauling of infrastructure, which will be carried out by private enterprise and government agencies. Mass transportation and private vehicles will require a total makeover, probably several generations of makeovers. The demand for computers and software will explode, as computerized guidance systems and traffic-control are installed to accommodate a mix of preprogrammed air traffic and free-flying vehicles moving in a broad spectrum of speeds and altitudes, ranging from hovercraft to space vehicles.

As a consequence of increased speed and ease of transportation, the farthest corners of the world will become neighbors: Surfers will be able to catch a morning wave in New Zealand and an evening curl at Malibu. Businessmen will be able to attend meetings in Moscow, Paris, and New York in a single day. Office workers will be able to commute to San Francisco from a cabin in the Sierras or a suburb of Denver. Soon the whole world will begin to resemble Alaska, where light planes are currently the major mode of private transportation. But there will be differences: Gravity control vehicles will be able to take off vertically in any weather conditions, making the whole world a runway, and they will fly far faster than conventional aircraft, including the fastest jets in the skies today.

Freed from gravity and inertia, architecture and construction will run wild—in the vertical *and* horizontal dimensions. New forms of buildings and private dwellings will sprout like mushrooms, not only in cities, but in areas such as islands and mountaintops that have never before been considered viable housing sites. Bora Bora will become an address like any other.

By 2024, the world will be experiencing a frontier movement not seen since the Wild West era in the United States. The mid-1800s was a perfect storm of seemingly unlimited territorial expansion, the Gold Rush, and powerful new technology, such as the steam engine, transcontinental railroad, mass manufacturing, and the increasing speed and freight carrying ability of oceangoing vessels. In 2024, space and the formerly remote corners of the world will be the new frontier, gravity-control vehicles will be the covered wagons and steam engines, and giant space stations and lunar or Martian colonies will be the land-rush destinations, while the asteroids, outer moons, and planets will be the new Sutter's Mill, Virginia City, and the Klondike. Fresh water can even be mined from Ceres or Io and brought back to Earth or the space colonies. Energy itself will become a hot new mining prospect.[3]

SOCIETY AND CULTURE

As hinted in the scenario above and *Chapter 4*, p. 13, the conquest of space has a spiritual dimension. It is impossible to predict what its effects will be, since space travel by human beings on a mass scale is an event without precedent, but astronauts have all reported a humbling sense of awe when viewing the Earth from space. Author Frank White coined the term "overview effect" to describe it, and predicts that as it spreads it will become increasingly difficult to sustain the tribal and national conflicts that rely on an "us and them" mentality.

Societal conflicts are fed not only by suspicion of "the other," but by territoriality and a sense of limited resources. The opening of the space frontier will undermine the notion of scarcity, which models life as a zero-sum game, where every winner has a matching loser. This encourages cutthroat competition, wealth-hoarding, and austerity, which in turn aggravates tensions between nations, religious groups, and the haves and have-nots. The new technology will not reward behavior of this type. Rather, freedom from gravity and inertia will create a sense of wonder at a whole new world of technological toys and experiences, as well as a renewed faith in science and the power of the human mind to overcome all barriers. In effect, it will do for the physical world what computers and the Internet have done for the world of communication: annihilating the barriers of time and space and language that we take for granted today. There will be a feeling of cutting loose from ties to the past and looking forward to a wide-open future. It will also stimulate a wave of optimism, inventiveness, entrepreneurial spirit, individualism, and social change. In the absence of reinforcement, the scarcity mode of thought and its associated negative behaviors are likely to loosen their grip on the body politic. Leaders who embody these values will find it more and more difficult to hold office and will adapt.

Gravity and inertia have always limited how far we can go, how fast we can get there, and what we can take with us. Gravity control means freedom for humanity on a scale never before known. Free from limitations of gravity and inertia, what kind of cities would we build, and where? How far away would New York, London, Paris, Moscow, Bali, Tokyo, or Peru seem to businessmen and tourists when they are all as close as the nearest 5,000-mile-per-hour airbus? Where would you go for a weekend adventure when highways on the ground became highways in the sky? How far would you live from work when travel between San Francisco and New York shrinks to an hour and you can land on a parking lot in

---

[3] Current technology would allow the conversion of solar energy into microwaves which could be converted back into electricity on Earth. Read the award-winning proposal of the interdisciplinary Space Solar Power D3 team at http://www.nss.org/news/releases/NSS_Release_20160307_SSPD3.html

the middle of the city? How about building colonies on the Moon? Terra-forming Mars? Mining the asteroids? Constructing factories and cities in space? What if we could tow multimillion-gallon tanks of water from Europa to feed domed cities on a terraformed Mars or irrigate the agricultural valleys and plains of California and the Midwest? What manufacturing processes and entertainment attractions would be possible in zero-gravity environments, not only in space, but here on Earth?

All of these ideas will move off the pages of a sci-fi novel into reality if one small experiment is a success.

Chapter 6

Big Oil

THE COMPANIES comprising "Big Oil"—including the automotive industry, which is treated in the next chapter—would be well-advised not only to *welcome* gravity control, but to zealously pursue its development. Why? Monster profits and a single, magical solution to all their most vexing problems. In addition, the new technology and the old have much in common. The leap from one to the other will pay for itself as fast as the market grows. Big Oil need lose nothing; all they need to do is ride the wave into the future and leave their current problems behind.

Today the petrochemical industry faces an existential crisis, like the whale oil industry that it replaced over a century ago. Gravity control or any of a host of alternative energy sources have nothing to do with the crisis, which has been building for decades. The threat comes from within, for if we continue to produce and consume fossil fuel as we have done in the past, the world will end, and that's a terrible thing to do to a customer base. Oil company executives will share the same fate as everyone else in this scenario, a notion that doesn't seem to trouble them at all right now. But once again, Upton Sinclair said it better than me: "It is difficult to get a man to understand something, when his salary depends upon his not understanding it." (It's safe to assume that this statement applies to women, too.)

The "Upton Sinclair method" of coping with the crisis is to deny that it exists and keep on truckin' as long as possible, which in the case of fossil fuel is something like careening down a mountain road in an eighteen-wheeler with one's eyes closed and one's fingers in one's ears. That seems to have been the strategy of Big Oil to date. But let's be fair: *What else can be done?* Big Oil is relentlessly encouraged by the sheer magnitude of their enterprise and the profits and power it generates. All this is *real*, and it's *today*. If the problem of climate change haunts them at all, they feel as trapped as anyone else, *because there's no way out*. There is no alternative to fossil fuel other than energy sources that put people inside the industry out of work and cut into company profits.

Until now.

What Big Oil *must* see is that gravity control is completely unlike solar, wind, hydroelectric or geothermal power. Gravity control is *not* green energy, it is a gusher of green *money*. It is *not* a competitive energy source—rather, it offers a way to escape the inevitability of peak oil, diminishing ROI, regulatory interference, an increasingly hostile public, multibillion-dollar lawsuits and monster settlements for environmental catastrophes, not to mention the slings and arrows of Mideast politics. All of these trends converge on a bleak future for the oil industry, where it's getting harder and harder to extract less and less profit. There might be a momentary burst of profit as oil becomes scarce, but then what? Gravity control opens the door to an alternative future in which the major energy corporations not only survive, they diversify and flourish for a thousand years, far longer than they will if current trends remain unchanged.

First, let's play dumb and ask a basic question: What is the purpose of companies such as BP, ExxonMobil, and ConocoPhillips? Is it providing oil, coal, and gas to a hydrocarbon-hungry world?

No, it is not. Though appearances might make us think so, this kind of thinking is a trap.

Is it providing energy, then? BP's new slogan—"Beyond petroleum"—might indicate something like that, but Big Oil doesn't *behave* that way. But let's give credit where credit is due: BP's slogan is encouraging in that it indicates a willingness to "think different." "Thinking different" is essential if the major oil companies are to survive.

While "Providing energy is our business" is an agreeable idea, the CEOs and board members of all the major petrochemical corporations will probably agree that their companies really have only *one* purpose, and that is *to make a profit for the shareholders of the company.*

That's not some kind of big whoop. It's just not anything that corporations like to boast about, except perhaps in shareholders meetings and quarterly reports. But there's more than a lust for money behind this purpose. It's the *law.* Failing in the duty to maximize profit can leave directors and officers of the company open to being sued by shareholders.

So *making a profit* is the prime directive. And if the company could accomplish the prime directive without producing any oil at all, not one shareholder would shed a tear. As we'll see, this hypothetical proposition is not far from the reality offered by gravity control.

Proof? Hard facts?

Okay.

On July 18th, 2015, a fast-rotating, one-kilometer-long asteroid called "2011 UW158" passed within 1.5 million miles of Earth—close by astronomical standards. Spectroscopic analysis revealed that this tumbling space rock was mostly *platinum.* The price of platinum today is roughly $1,000 per *ounce,* or $30,000 per kilo.

If BP, ExxonMobil, Royal Dutch Shell, or ConocoPhillips were in the space-mining business instead of the riskier, costlier, and less profitable oil business, one or more of them would have bagged an easy $300 billion to $5 trillion from this *one* asteroid. And the solar system is *full* of asteroids. Where would the market be for that much platinum? The supply would inundate today's market, diluting profits, but *tomorrow's* market will be much bigger, because the new industrial economy of gravity control is going to dwarf the old. And Big Oil can lock in a sizeable share of the boom by diversifying its holdings today, thereby decreasing risk tomorrow.

To put things in perspective, the value of asteroid 2011 UW158—one asteroid—exceeds that of the world's annual production of coal, oil, and gas *combined*. But there's more. In the oil business, companies prize concentrated reserves that are easy to discover and easy to extract. Here, the entire reserve was discoverable by telescope (no costly exploration), and the entirety of it was concentrated in a one-kilometer package, ripe for the picking.

"But we're in the *oil* business, son—not the space business," say my Texas friends.

Well, I used to live in Houston, so I know where you're coming from. However, I feel obliged to remind you that back on earth, oil profits are *down*. According to the New York Times, Royal Dutch Shell's profit fell 56 percent in the fourth quarter of 2015 over the same period a year before: $1.8 billion, down from $4.2 billion during the same period in 2014. When Shell Oil abandoned Arctic drilling efforts recently, the reason had nothing to do with protesters who attempted to block their ships from leaving port. Rather, drilling in the Arctic Ocean for $90-a-barrel oil had become unprofitable in a $50-per-barrel market.

Now let's do an experiment: Put $5 *trillion* in one hand and $1.8 billion in the other, and then tell us which hand you like better.

While you're thinking about that, let me point out that there are more parallels between the space business and the oil business than it would seem at first.

An oil company sends scientists on costly exploratory expeditions to distant corners of the globe to probe deep under treacherous oceans in search of hidden natural resources. And when they find the buried treasure, they send ships and heavy equipment out after it. Substitute "spaceships" for oceangoing ships, and everything stays the same in the space-based resources business. What's hanging us up now is the $12,000-per-pound cost of space travel. With gravity control, however, that price will drop to less than the cost of an oceangoing exploratory mission. Rewards will go up, cost will go down.

Back to oil: The next step is to set up a drilling platform where workers will live for days, weeks, or months away from civilization. Divers don airtight diving suits—like spacesuits, right?—and plunge into a weightless environment where one false step means death. When you consider that in some corners of the globe, oil company workers are attacked by hostile "aliens," the parallels between the oil business and science-fiction become rather striking.

Platinum is far from the only resource in space. Aluminum is another, and the demand for aluminum in the new economy will be enormous, as it is the principal element in GCV hulls. But one of the most valuable and abundant commodities is "the oil of the future": *water*, which has been found on Ceres (a dwarf planet in the asteroid belt), and several moons of Jupiter and Saturn. Saturn's famous rings are mostly *water ice*, probably expelled from a global ocean on Saturn's moon, Enceladus. The volume of water on Jupiter's moon, Europa, *alone* is two to three times greater than all the water in all the oceans of the Earth! Even Pluto seems to have oceans of pure, four-billion-year-old water.

At this point you might say that the demand for oil is well-established and the demand for space commodities isn't. That's certainly true—*today*. But one must look at space-based industry the way John D. Rockefeller looked at Ohio oil in the 1800s: *as a colossal opportunity*. Space exploration is coming with or without gravity control, but slowly. *With* gravity control, it will be *galloping* at us tomorrow. Meanwhile, oil resources are dwindling, just as whale oil was in Rockefeller's time.

The fact that Big Oil is not in the space business today is irrelevant. Companies and agencies such as NASA, SpaceX, and Planetary Resources—which *have* staked a claim in space exploration and space-based business—had to sit on their hands and watch asteroid 2011 UW158 cartwheel by in July, 2015, because they didn't have the technology to intercept it and extract the treasure. Gravity-control technology does several things that are impossible with present technology: First, it brings space exploration costs into a range on a par with oil exploration costs. Second, it enables spaceships the luxury of maneuvering in deep space without consuming prodigious amounts of costly fuel. This in turn makes it possible for a space vehicle to dock with a cartwheeling mountain of platinum, extract tons of booty, and take it back to Earth or the nearest space station, colony, or space manufacturing site for processing. There's no reason an oil company couldn't build and own those plants, too, and open them up to a lucrative tourist trade on the side, by the way.

When gravity-control arrives, companies with vast resources, such as Big Oil, will be better positioned than any other to diversify into space-based mining and other industries. This is why Big Oil needs to jump on the opportunity *now*, or else wind up making expensive leasing contracts with companies that *are* first to build gravity-control empires and space-based mining stations (like the oil platforms of today). If Big Oil were to make a small initial investment today, their "sci-fi side business" would quickly overtake their main profitmaking enterprises, and they would secure the patents that will be worth trillions tomorrow. Early investment is the way to improve the company's profit picture during the transition. To *wait* is to *lose* this historic opportunity *forever*.

The moral of the story of asteroid 2011 UW158 is clear: *Space is overflowing with material resources and opportunities for private enterprise.* Asteroid 2011 UW158 is but one nugget in a sea of gold. And the game-changing technology described in this book throws open the doors to the new Gold Rush, *Platinum* Rush, as the case may be.

The cost of a single offshore drilling rig is $200 million. The cost of redoing the 1994 experiment is a thousandth of that—roughly the price of a three-bedroom, two-bath middle-class house. A cheap investment today that reaps *trillions* of dollars tomorrow? What is there to lose? The alternative is to wait and be blindsided, like the record companies who were *shocked* when digital file-sharing "snuck up on them" in plain sight.

This chapter was written to show the petrochemical industry how to avoid the pain and go directly to the profit by the shortest possible route. As the record industry learned, failure to adapt is more than a pointless exercise in futility, it's a colossal waste of money. The future of Big Oil is out there, and if you squint your eyes a bit, you can see it—in the form of a gravity-control air-car.

Chapter 7

The Automotive Industry

THE ADVENT of gravity control means the realization of a long-held dream for the auto industry: the merger of flight technology with private transportation. The product of this merger will be an explosion in sales beyond the wildest dreams of the world's largest automakers today.

The industry is already diluting the gasoline-driven market with electric cars and hybrids. The introduction of an all-electric, gravity-control vehicle merely extends this trend into the vertical dimension. Much as it was with the "horseless carriage," lack of familiarity with gravity control will rapidly fade as friends, families, and relations of early adopters begin to enjoy the spectacular and exhilarating sense of freedom afforded by the new line of vehicles.

Past attempts to build a light plane that can fit in a garage and function as a commuter vehicle have failed to measure up to the needs of an average family with a three-bedroom, two-bath house, 2.5 children and a thirty-year mortgage. Barriers include:

- Price
- Pilot skill (and eyesight)
- Weather vulnerability
- Infrastructure limitations; i.e., the awkward wingspan of a light plane, the need for landing fields, hangars, air traffic towers, and ground crews (*cont. next page*)
- Limited load capacity
- Safety considerations

Gravity control solves all of these problems in a big way:

Price: Cost-savings are inherent in GCV design. They have few moving parts and do not require powerful jets or propellers. Neither do they consume much fuel. The principal material used in their construction is aluminum, one of the most abundant metals in the Earth's crust. As long as the vehicle operates below 14,000 feet, a pressurized cabin is unnecessary. Expenses will concentrate in computer and sensor systems, the power plant, and the gravity-control generator. Mass production will bring all of these within range for a middle-class income, but ownership is not a requirement. Robotic, government-owned private GCVs can ferry people to and from work on a pay-per-flight basis. Mass-transit systems are also possible.

Pilot skill: If average people were required to learn how to pilot a high-performance gravity-control vehicle, the barriers would be insurmountable. Fortunately, there will be no such requirement. According to *Business Insider*, companies such as Mercedes, BMW, and Tesla have already released, or are soon to release, cars with some ability to drive themselves. High-tech companies, such as Google, are also getting into the act with completely driverless vehicles. *Business Insider Intelligence* predicts that ten million self-driving cars will be on the road by 2020. The point is twofold: first, customers will soon be accustomed to riding in self-piloting vehicles; second, the AI software already being developed for self-driving vehicles will be adaptable to gravity-control vehicles. The pilot of a gravity-control vehicle will punch in a destination, sit back, and let the GCV take over. (See *Chapter 5,* p. 19, for a detailed scenario.) Human error will be sharply reduced with autopilot GCVs.

Weather dependency: Flying in a light plane can be bumpy, even in calm weather. In rough weather, the gyrations become violent and frightening. Bad weather or poor visibility can also preclude flying altogether. Gravity-control aircraft are non-aerodynamic, so weather conditions are almost irrelevant. Stability can be achieved several ways, including computerized adjustments according to homing beacons, or the Earth's magnetic field. Since gravity control also reduces inertia, passengers will never experience a bumpy flight, no matter what gyrations the craft undergoes.

Infrastructure limitations: Light planes require runways for takeoffs and landings and hangars for storage, which virtually excludes them as commuter vehicles for someone with an average income. Gravity-control vehicles will be capable of vertical takeoffs and landings in any weather, at any speed, from virtually any flat surface (a backyard lawn, for example), which means they are fully adaptable to urban and suburban environments and do not require airfields.

Since airfields will not be needed, there will be no need for secondary ground transportation from the airfield to the actual destination and from the destination back to the airfield. Personal GCVs will be door-to-door vehicles, which, like automobiles, can be stored in parking garages between flights or simply parked in open lots surrounding destination buildings. But GCVs differ from automobiles in two important respects: they do not require roads, and they can easily fly five times as fast as an automobile between destinations, traveling *in a straight line.*

Load capacity: Like automobiles, private GCVs can be engineered for two, three, four, six, or more passengers. As the number of passengers increases, so does cabin size and disc or sphere diameter. Optimal configurations must be determined by research and development.

Safety: Planes, trains, and automobiles have safety risks associated with them, including collisions, high-speed machinery, engine heat, explosive fuel and cargo, and human error. GCVs are highly maneuverable, so collisions are less of a concern. They have no high-speed machinery associated with them (other than the GCV in flight). Problems may arise from heat generated by the power plant and the effects of dynamic nuclear orientation on biological organisms, but the problems are soluble. The saucer shape is aerodynamically suited to high-speed flying within our atmosphere. The gravity-control process will probably create a layer of ionized gases around the hull of the craft, which will allow it to travel at supersonic speeds without significant friction with the atmosphere or sonic booms.

Microwave radiation presents safety problems. The plan is to contain the generator in a shell at the center of the GCV below the cabin and conduct pulsed dynamic nuclear orientation from the generator to the hull of the craft by aluminum spokes with colloidal magnesium or iron inclusions to assist penetration of the effect. Passengers can be shielded from exposure by cabin walls, helmets, and flight suits made with aluminum fiber. There may also be ways to induce the gravity-control effect *without* the use of microwaves. One such method involves an aluminum flywheel. My father was experimenting with this at the time of his death.

DESIGN FEATURES

In order to reduce costs while enhancing user experience, terrestrial gravity-control vehicles will be less sophisticated than spaceworthy models. A class of vehicles with reduced flight characteristics will emerge to appeal to users who are comfortable with automobiles. These would resemble the "land speeders" that appear in the *Star Wars* films and would allow limited use of manual controls.

A present-day hovercraft—Paul Moller's M200G Volantor—looks much like the envisioned GCV air car. The M200G is shaped like a saucer, with eight computer-controlled fans around the perimeter, a domed cockpit made of clear plastic in the center, and a stabilizer fin in the rear. Moller's fans are powered by gasoline or diesel fuel. At present, the M200G has a load capacity of 350 pounds and a computer-regulated operating altitude of ten feet in order to comply with FAA regulations (if its altitude were to exceed ten feet, it would be regulated like an airplane). The M200G can fly over any kind of terrain at 50 mph.

A gravity-control saucer would have *electric* fans a fraction of the size of Moller's current fans, as minimal thrust goes a long way in a vehicle with an effective weight of a few ounces at most. Compressed gas jets have also been suggested. The stabilizer fin in the back would be reduced or eliminated in order to minimize drag; it's only purpose in a small saucer would be to keep the vehicle oriented "forward" in flight. Some lift comes from the fans or jets. The rest comes from buoyancy, as the craft becomes lighter than air when the gravity-control effect reaches maximum.

The top speed of propeller-driven aircraft *without* gravity control is 670 mph. We can assume that *with* gravity control, it would be much faster. Air resistance is a limiting factor, but as noted above, it is anticipated that the GCV will be surrounded with a "skin" of ionized gases that will enable it to slip more easily through the atmosphere.

Other kinds of propulsion, including turbojets, ramjets, and scramjets, will enable supersonic or hypersonic speeds. Turbojets are attractive because they can be fueled with alcohol. Bear in mind that these jets would be tightly engineered, but *miniature* by the standards of today, on the order of model airplane jets. The ultimate form of propulsion would be an electromagnetic drive that utilized the Earth's magnetic field. The energy in the field is 0.25 to 0.65 gauss at the Earth's surface, which is minuscule, but more than enough to move a "weightless" compass needle. A vehicle under the influence of gravity control will weigh *less* than a compass needle, and powerful electromagnets onboard can be directed at any angle to the Earth's magnetic field, enabling astounding mobility and acceleration without burning any fuel or making any sound. A sophisticated drive system such as this would be ideal for space travel. Once the vehicle has exited earth's gravitational field, it can use the gravitational fields of celestial objects to maneuver, simply by increasing the gravitational force on one side of the craft and decreasing it on the other. While solar power would be helpful, such vehicles will probably require some kind of onboard reactor. For terrestrial vehicles, turbojets and propellers are a far more economical and attractive alternative. A slotted rim on the saucer spinning at high speed has also been proposed.+-

Most of the vehicle's energy would be invested in maintaining the gravity-control effect. A capacitor in the belly of the craft is suggested for cycling the current to the microwave generator. The weight of the capacitor, which will account for most of the rest-weight of the vehicle, is almost irrelevant, because it is neutralized in the preflight warmup, along with the weight of the occupants. If such a system can store enough power to maintain the gravity control effect for three hours, the range of the vehicle would be most of the North American continent. There would be no need to "recharge." The vehicle would simply touch down at a "fueling station" and exchange a dead battery for a live one. The fueling station would attach the dead battery to a charger and exchange it out for the next customer. Alternatively, air scoops could drive turbines attached to a charger.

A spaceworthy vehicle will cost more than a terrestrial craft. The added cost stems from the need for added shielding, structural strength, advanced instrumentation for navigation, communication, and viewing, added power to feed an electromagnetic or gravitational drive, and backup propulsion systems for maneuvering near planets without an electromagnetic field.

In order to preserve the integrity of the gravity-control effect, the dome of a terrestrial craft should be covered with a semi-transparent aluminum film. Advances in material science should yield increasingly elegant solutions to this problem, such as transparent metal.[4] The GCV will be equipped with a

---

[4] Aluminum oxynitride is a promising material, since it is strong, 80% transparent, and aluminum is a principal component.

parachute in case of emergency. Drivers will need simulation training to prepare for the eventuality of a bailout, but a breakdown while airborne is less likely to occur in a gravity-control vehicle than a conventional aircraft. GCVs experiencing heating problems can simply touch down almost anywhere: no landing strip required. GCVs will get an automatic safety check after each landing. Those who could not adjust to the idea of a high-altitude bailout could use vehicles that maintain a ceiling of one to ten feet, like the Moller M200G mentioned above.

CONCLUSION

Gravity control means the long-sought merger between private transportation and air travel. This will touch off a chain reaction of economic growth that is not only unprecedented, but *unfathomable* to us now. A similar situation existed in the mid-1700s at the dawn of the Industrial Age, in the late 1800s, at the beginning of the automotive age, in 1903, when the Wright Brothers flew at Kitty Hawk, at the end of the 1940s, when jet aircraft superseded propeller-driven flight, and in 1970, at the beginning of the computer age. We might even include the year 2007, when Apple put the first smartphone on the market. No-one alive at the dawn of these eras sensed all that was implied by these technological innovations, and in our opinion, the changes in store for society in general and the automotive industry in particular because of gravity control far outstrip *any* of the above. We would go so far as to say that the world is about to undergo an economic "leap to hyperspace."[5]

The revamping of transportation will include a thousand-and-one new vehicle designs, ranging from those with performance characteristics not unlike today's ground-based transportation up to and including private space vehicles. New manufacturing sites will be needed to produce these vehicles, which presents an opportunity to reinvigorate Detroit and other depressed manufacturing areas. Eventually much of the manufacturing will move off-planet, which has a host of implications worthy of a book in itself. All sectors of the industry will be affected, including trains, trucking, public and private transportation, and transcontinental shipping. The new field of space commerce will require countless new vehicle configurations, all of which would be science-fiction by today's standards.

Ancillary industries will receive a shot of adrenaline from the economic activity unleashed by gravity-control technology. All new vehicles will require new infrastructure, new kinds of support and repair, and new onboard electronics, including sophisticated new computer systems for navigation and flight control. As people abandon four wheels for "no wheels," the materials in existing automobiles will be recycled into GCV hulls and other uses. Rather than go on listing the effects of gravity control on the automotive industry and other spheres of economic activity, I have created a scenario of the world in 2057 A.D. in *Chapter 5 – A Trillion-Dollar Technology* (see p. 19).

---

[5] See *Star Wars – Episode IV*, if the terminology is unfamiliar.

Chapter 8

Stakeholders

ASSUMING the attempt to replicate the 1994 experiment meets with success, the companies listed below are in line to benefit. Military contracts will figure large in the future of gravity control, but here we are primarily concerned with private sector growth.

Currently, gravity control is surrounded with a quasi-mythical aura, which disguises its forthcoming role as an integral part of everyday life. The purpose of this list is to dramatize the public role of the technology. The list is far from complete, but it suggests avenues for mainstreaming.

One important point emerged while composing the list: Gravity control will break down boundaries and categories between companies. For example, the boundary between spacecraft and aircraft will blur, as will the boundaries between automobile manufacturers and aviation manufacturers, self-driving earthbound cars and self-piloting GCVs, spectroscopy and propulsion, software and avionics, helicopter makers and aircraft or automobile makers, drone companies and private aircraft or space exploration companies. Every company on this list should look at the potential for creative mergers in the near future.

Companies that invest in R&D now will reap greater rewards in the future, while companies that prefer to wait for someone else to "prove it out" will have to get in line to pay licensing fees.[6] Companies in categories (2) through (8) below should consider mergers.

1) MICROWAVE, RADAR, AND SPECTROSCOPY COMPANIES

Delphi, EEC, JEOL, Microwave Dynamics, Microwave Solutions Inc., Zeiss, and other companies too numerous to mention.

2) ELECTRIC VEHICLE MANUFACTURERS

BMW, Chevrolet, Daimler (Mercedes-Benz), Fiat, Ford, Kia, Nissan, Tesla, Toyota, Volkswagen

---

[6] Company representatives email info@klafraknar.com. Include verifiable contact information.

3) **AIR CARS**

Aero-X, Hoverbike, Moller Skycar, Terrafugia TF-X, X-Hawk

4) **LIGHT AND ULTRALIGHT PLANE MANUFACTURERS**

Aero Adventure, Aeropro CZ, Airdrome Aeroplanes, Arion Aircraft, Carlson Aircraft, Cessna, CGS Aviation, Dynali Helicopter, Earth Star Aircraft Inc., Fisher Flying Products, FK-Lightplanes, Flightstar, Hipps Superbirds, ICON Aircraft, IndUS Aviation Complex, Jabiru USA Sport Aircraft, Kappa Aircraft, Lockwood Aviation, Loehle Aircraft, New Kolb Aircraft Co, Nexaer, Oregon Aircraft Design, Piper Aircraft, Preceptor Aircraft Corp., Quad City Ultralight Aircraft Corp., Quicksilver Manufacturing, Remos Aircraft GmbH, Sonex Aircraft, Tecnam Aircraft

5) **CORPORATE JET MANUFACTURERS**

Boeing Business Jets, Bombardier Business Jets, Citation, Cirrus Aircraft, Daher-Socata, Diamond Aircraft, Eclipse Aerospace, Embraer, Extra Aircraft, Gulfstream Aerospace, Hawker Beechcraft, Honda Aircraft, Learjet, Piaggio Aero Industries, Pilatus Business Aircraft, Piper Aircraft, Quest Aircraft, Syberjet Aircraft

6) **MAJOR AIRCRAFT COMPANIES**

Airbus, Boeing, Bombardier Aerospace, Cessna Aircraft, Dassault Falcon, Lockheed Martin, Piaggio Aero Industries

7) **HELICOPTER COMPANIES**

AgustaWestland, Bell Helicopter, Enstrom Helicopter, Eurocopter, MD Helicopters, Robinson Helicopter, Sikorsky Aircraft

8) **DRONE MANUFACTURERS**

Aerovironment, Amazon.com, Boeing, Dajiang Innovation, Denel Dynamics, Lockheed Martin, Parrot EPA, Facebook, Google, Matternet, Skycatch

9) **COMPUTER COMPANIES, SOFTWARE DEVELOPERS**

Airware, Apple, Microsoft (for development of local, national, and global interactive flight control networks and security), Avionics software developers such as ENSCO, Aurora, Cobham (for vehicle control and safety)

10) CIVILIAN SPACE EXPLORATION

The line between the following companies and conventional aircraft companies is about to blur.

Bigelow Aerospace, Blue Origin, Ford Aerospace, Orbital Sciences, Planetary Resources, Scaled Composites, Space Adventures, SpaceDev, SpaceX, Stratolaunch Systems, Virgin Galactic

11) MANUFACTURING AND CONSTRUCTION

Gravity control allows duplication of space manufacturing environments in terrestrial laboratories. Uses include crystal growth, processing of immiscible alloys, formation of spheres from molten metal, and medical/bioengineering applications.

Pulsed nuclear orientation can be conducted to robotic equipment on factory floors via aluminum wires, reducing inertia in high-speed equipment. This will allow robot armatures to run at extremely high speeds with low energy consumption and without overbanking or tearing the machinery apart.

With gravity control, it will become possible for small crews to position or move large masses without the use of cranes.

ABB Ltd., Adept Technology, Apex Automation and Robotics, Aurotek, Axium, Denso Wave, Ellison Technologies, Kawasaki Robotics, Kuka AG, Fanuc, Nachi Fujikoshi, Pari Robotics, Reis Robotics, Rockwell Automation, Schunk GmbH, Staubli International, TM (Toshiba Machine) Robotics, Yamaha Robotics, Yaskawa Electric

12) PETROCHEMICAL COMPANIES

See *Chapter 6, Big Oil,* for a discussion of the role gravity control will play in the future of petrochemical companies—*not* as a competitor, but as a new source of revenue and an eventual escape (without loss of profit) from industry-threatening problems, such as peak oil, decreasing ROI, increasing regulations and liability expenses for environmental destruction.

Major oil companies include Abu Dhabi National Oil, BP, Chevron, ConocoPhillips, Eni (Italian multinational), ExxonMobil, Gazprom (Russia), Iraqi Oil Ministry, Koch Industries, Kuwait Petroleum, Lukoil (Russia), National Iranian Oil Co., Nigerian National Petroleum, Pemex (Mexico), Petrobras (Brazil), PetroChina, Petroleos de Venezuela, Petronas (Malaysia), Qatar Petroleum, Rosneft (Russia), Royal Dutch Shell, Saudi Aramco, Sinopec (China), Sonatrach (Algeria), Statoil (Norway), and Total (France).

13) TRAVEL AND ENTERTAINMENT

Eventually gravity control will be incorporated into theme-park rides and exhibits.

Magicians who adopt now, before gravity control is an everyday fact of life, will be able to perform astounding stunts while changing the course of history by helping to mainstream the world-changing technology. (Email info@klafraknar.com and include verifiable contact information.)

The travel industry will undergo the greatest transformation of all. Transportation costs will drop and the globe will shrink as privately owned GCVs routinely begin to fly at Mach 2 or more. Since takeoffs and landings will be vertical, airports will no longer require runways and taxiing areas. Landings can take place on remote islands and mountaintops without airstrips. More important, space itself will become a destination, with sightseeing tours at first, then sojourns on space stations or lunar and Martian colonies later.

Chapter 9

Geopolitics

THIS CHAPTER may well be moot, since the science in this book has yet to be proven conclusively, and a rerun of the 1994 experiment may show that the results were in error. However, if we are willing to entertain the notion that gravity control is real, the question arises as to whether it should be released now, when international tensions are high and the established order is ailing.

The answer is "Yes, now more than ever." But the question implies that this book contains some new revelation when the opposite is true. A complete starter kit for the new technology was presented in 1981. Every effort was made to initiate a research program between 1981 and 2015, and the powers that be, including the defense industry, never showed the slightest interest. The science has remained on the public record for almost forty years. What makes this book new is the intensifying threat of climate change. And that's the point: If your house is on fire, you *want* someone to haul you out of bed and help you put it out, preferably with a firehose, not a water bucket.

Had gravity control been mainstreamed in 1981, climate change might be a thing of the past today, and the economic ills threatening the established order might never have arisen. Thanks to the delay, it may already be too late to reverse the damage. Therefore the time for waiting is long since *over*— the technology *must* be tested, and if it works, it must be mainstreamed *now*. Nothing offers a more effective way to influence climate change, while at the same time benefiting the global economy and providing the fossil-fuel industry a way to transition to something more profitable.

It's hard to think of anything that would trump the survival-of-the-species argument, but let's pursue the question a little further:

When has the world *ever* been free from turmoil? And when has turmoil or the threat of it ever stopped a new technology from making its debut? The A-bomb debuted *during* World War II, not before. If it were necessary for world peace to break out before releasing a new technology, we'd still be running around in animal skins, and when the sun set, we would retire to our caves to shiver in the dark while devouring raw meat, because after all, *fire* was one of the first technological innovations that might be used as a weapon.

Progress is always double-edged. In the majority of cases, we find that new technologies are absorbed by society, changing life for the better. And unlike fire, spears, or gunpowder, there is *nothing* inherently destructive about gravity control. Unlike nuclear fission, which can be used to make a bomb that can destroy a city or even all life on earth (which it may already have done, thanks to Fukushima), gravity control can't burn, can't explode, can't destroy anything at all. It *can* be used to explore the cosmos, build space-based factories, mine the asteroids, colonize the Moon and Mars, and bring fresh water back from the moons of Saturn and Jupiter. It can also give us the power to defend our planet from rogue asteroids or comets. *No more Chicxulubs!*[7] It will enable us to get rid of toxic waste once and for all and give the Earth a chance to heal. It will revitalize the economy and initiate a new Industrial Revolution. And it will do more to reverse global climate change than ten million windmills, by giving the major polluters something more profitable to do with their time and money. For these and other reasons, gravity control is nothing to fear: it is a force for *peace* and the survival of the species.

If it works, of course.

As noted above, gravity control technology was released at the 17th Joint Propulsion Conference in Colorado Springs in 1981. From there it promptly went into a state of suspended animation. Therefore the question isn't really whether it should be released now, but whether or not somebody has done something about it in the intervening time. Scientists from every country in the world have had unhindered access to the 1981 paper and several other related publications for thirty-five years. Even the success of the experiments has been alluded to in at least four publications, including a peer-reviewed paper. So there is no odor of Edward Snowden about this book. More like Paul Revere.

Instead of worrying about the effect of *Gravity Control with Present Technology*, we should be concerned with whether or not the 1981 paper has already been researched by foreign military powers, a segment of society not known for broadcasting their secrets. If this is the case, *and* the technology works, it will create exactly the imbalance of power so dreaded on all sides. As occurred with Sputnik in the 1950s, the U.S.—which has been blithely pinching pennies on *pure* research for decades—will wake up to find itself scrabbling to catch up to a technology one of our nation's *own scientists* tried to *give away* in 1981 and several times thereafter.

It is my hope that the unpleasant prospect of the People's Republic, the Russian Federation, or any number of small universities in Germany, Eastern Europe, Scandinavia, or the Middle East laying their hands on gravity control—whether in the past, present, or future—will prompt the institutions that ignored my dad's proposals between 1960 through 2012 to reconsider.

Again, it must be emphasized that the map to gravity control has been published five times over the

---

[7] The asteroid that ended the age of the dinosaurs struck near Chicxulub on the Yucatan peninsula. Six miles in diameter and traveling at 50,000 mph when it hit, it exploded with the kinetic energy of 300 million nuclear bombs. Currently we have no reliable means to avert an extinction event such as Chicxulub.

last thirty-five years prior to this book, with easy access for all. My concern is that the U.S. may have dropped the ball, allowing some other country to get a lead in research and development.

Now let's take a deep breath and consider the whole thing from another angle. You know that deep breath you just took? It's got 400 parts per million of CO_2 in it, and rising. The safety ceiling for atmospheric CO_2 is 350 parts per million. Earth's CO_2 count has been higher than that *since 1988* and it is rising *at an accelerating rate*. "Safety" is a concern because CO_2 is a greenhouse gas, and manmade CO_2 is the number-one cause of global warming.

Did I mention methane? Not the often-mocked methane from cow farts, but the *trillions of tons* of methane trapped on the ocean floor and in the Arctic tundra. It's beginning to bubble up now, thanks to the warming ocean currents brought on by the buildup of greenhouse gases. Methane is ten times more effective than CO_2 at trapping heat in the atmosphere. While I was drafting and revising *Gravity Control with Present Technology,* more than 91,000 metric tons of methane leaked from one storage facility in southern California, the equivalent of burning 862 million gallons of gasoline. But this little burp is nothing compared to what is about to belch from the ocean floor in the Arctic.

The global warming cycle is self-augmenting, like feedback in a guitar amplifier. The worst that an amplifier with feedback can do is blow out your eardrums. But CO_2 and methane can extinguish all life on Earth—not just human life, but *all* life.

Does that sound like a major geopolitical concern? Does it sound like something that might provoke wars? Upset the financial markets? Ruin Christmas?

Senator James Inhofe (R – Oklahoma), author of *The Greatest Hoax: How the Global Warming Conspiracy Threatens Your Future*, might not think so, but the Pentagon does. Google "Pentagon global climate change report" and click on the link to the July 29, 2015, Department of Defense report (612710).

In 1981, the year of the Colorado Springs paper, most people in the U.S. thought of global climate change as a threat to beachfront real estate values in Florida. That was all. We did not understand the self-augmenting interconnectivity of climate factors. When my father began writing research proposals a few months after delivering the Colorado Springs paper, he put in a few sentences about how the technology would alleviate global warming, but he removed them from the final draft. "Nobody believes it," he told me. "They'll think that I'm an alarmist." He was having enough trouble selling "antigravity" without mentioning "global warming" in the same breath.

I, on the other hand, freely admit to being an alarmist. And I'm in good company. The website skepticalscience.com cites a study that shows a 97% consensus on human-caused global warming among publishing climate scientists, and the higher the level of expertise, the higher the consensus. [8]

If gravity-control technology proves to be real, then the failure to bring it online in 1981 has to rank as one of the most colossal bureaucratic and scientific blunders in human history, though no one may be around to care about it in a couple of hundred years. It is much the same as if the Einstein-Szilárd letter of 1939 (concerning the possibility of the Nazis building an atomic bomb) had been ignored by FDR and his advisors, many of whom called it "science-fiction." The result would have been the Axis getting the atomic bomb first, followed by the annihilation of London and Moscow, the Nazis winning World War II, and the unimpeded progress of "the Final Solution," which probably would have been extended to all "non-Aryans." As bad as this sounds, the effect of failing to mainstream gravity control could well be *worse*, because it might result in the extinction of all life on this planet.

If the technology had been mainstreamed in 1981, the environmental crisis we are facing today probably would never have arisen at all. Instead, we would be thirty-five years along the course of an alternative history characterized by an unprecedented economic boom, city-sized orbital space stations, space colonies on Mars and the Moon, and a GCV in every garage. Meanwhile our cities would be transforming into something out of *Star Wars* or *Star Trek*.

In the 1960 meeting described in *Chapter 12* (p. 71), an Air Force Colonel at Ames Research Laboratory told my father that an antigravity propulsion project would be classified higher than the Manhattan Project.[9] That made sense by Cold War standards. Now the pendulum has swung to the opposite extreme: In today's world, it is vital to *spread* gravity-control technology throughout all sectors of human society as rapidly as possible because it is the *only* way the technology can have a significant impact on climate change. The rewards far outweigh the risks. For those still inclined to worry about strategic imbalance, there are several reasons for reassurance.

First, as noted above, there is nothing inherently destructive about gravity control.

Second, it will be difficult to mount rockets, guns or bombs on a gravity-control vehicle. Maintaining nuclear orientation in spatial extent on a gunship or a bomber would require a high-output power source, such as an onboard nuclear reactor. The limited power reserves possible with present technology dictate seamless hull integrity. A gun turret or port or open bomb bay would cause the craft to plummet out of the sky. GCVs using conventional power sources *must* touch down before a hatch opens. Using a drone GCV to deliver a bomb is a possibility, but GCVs—especially those with weaker power plants—can be brought down with broadly targeted electronic countermeasures, such as a radar beam strong enough to disrupt the propulsion system, which means that a rogue craft using

---

[8] A PDF of the study is available at http://iopscience.iop.org/article/10.1088/1748-9326/11/4/048002.

[9] See *Appendix E*, p. 355

gravity control can be dropped from the sky at the speed of light *without* a direct hit by a missile.

Third, past is prologue: The evolution of gravity control will be much like the evolution of conventional aircraft and rocketry since Kitty Hawk. Airplanes have played an ever-increasing role in military operations, but on balance we accept this as the price we pay for the benefits they confer. The progress of GCV technology will be similar: Commercial benefits will accrue immediately, and design advances on one side will be met with advances on the other. In the end, progress will balance out just as it has with fixed-wing aircraft: the most sophisticated aircraft and spacecraft will wind up in the possession of the military, while smaller, lighter, slower versions will wind up in the hands of private companies and individuals. The power edge will always go to the military because societies always confer unlimited resources on national defense.

PROSPECTS FOR PEACE

The advent of gravity control will bring about profound changes in society. One of these might well be a decrease in geopolitical tension. While mankind seems to have an irrepressible urge to use every new technology as a weapon, there are features of gravity-control technology that may push humanity in the opposite direction, that is, toward peaceful cooperation, rather than violent competition.

First, the more the technology proliferates, the more human beings will see the Earth from space. The psychological effects of this experience are profound. In 1987, author Frank White coined the term "overview effect" to describe it in his book, *The Overview Effect: Space Exploration and Human Evolution* (a second edition was published by the AIAA in 1998). According to White, seeing the Earth from space left an indelible impression on all astronauts. From space, the conflicts that plague humanity seemed insignificant. In 1998, after returning from his final trip into space at age 77, astronaut John Glenn said, "To look out at this kind of creation and not believe in God is to me impossible... It just strengthens my faith." All seem to agree that it is impossible to contain the experience in mere words.

Within five years after gravity control goes mainstream, tens of thousands, perhaps more will have seen the Earth from space and experienced the overview effect. Within *ten* years, the number will be in the millions. Eventually hardly a living soul will fail to experience "the overview" at least once in a lifetime, while many will spend extended periods in space, living and working in space stations or colonies on the Moon and Mars. This will bring about lasting changes in human consciousness, and probably for the better. It is likely that the second or third generation born after the debut of gravity-control technology will look upon the conflicts of the twentieth and early twenty-first century as strange, barbaric, primitive, irrational, and incomprehensible from the new, cosmic perspective.

Second, much of the world's conflict over the last five thousand years has been predicated on competition over limited resources. Two or more groups want the same land, the same water, the same food supply or the same trade routes, and they go to war to wrest control from their enemies or to maintain control themselves. The assumption of scarcity drives the choice of leaders as well as policy. Remove scarcity and the whole structure will morph into something new. Gravity-control technology will catalyze this process. Resources in space are virtually limitless, and soon those resources will be coming back to Earth to energize commerce and industry.

But one need not go into space to see how gravity control will revolutionize resource allocation here on Earth. Among other things, channels of distribution will flow more quickly, building in remote areas (including space) will ease population pressure, and ease of construction will make living more comfortable for more people. In addition, employment will be at an all-time high, and traveling from one job-rich area to another will be more feasible for more people than ever.

CONCLUSIONS

• Global climate change, not gravity control technology, is the ultimate threat to geopolitical stability. Gravity control can help to reverse climate trends, but this will only occur if the technology saturates industrial societies as quickly as possible. The corporations now causing global climate change will be early adopters—not because they are altruistic, but because gravity control leads to higher profits while solving their most intractable problems (see p. 35).

• Concerns over national security are justified, but only if gravity-control technology is developed in secret by one country to the detriment of others. The existence of the technology has been an open secret for thirty-five years; this book merely promotes symmetrical development.

• The evolution of gravity control will parallel the evolution of aviation and aerospace, leading in an upward spiral toward strategic balance and cooperative ventures in space and on the ground.

• Widespread dissemination of the technology will have positive effects on society and may alleviate conflict, rather than exacerbate it. A short list of positive effects includes sustained and expanding economic growth, improvements in transportation and distribution, new frontiers for colonization and industry in space, proliferation of the overview effect, replacement of the culture of scarcity with the culture of abundance, alleviation of climate change, and the cleanup of toxic waste sites.

Part II

The History of an Idea

All great truths begin as blasphemies.

— George Bernard Shaw

Chapter 10

1954 – "Surely You're Joking, Dr. Alzofon"

THE STORY of gravity control might have begun in the 1940s at Cal, but we'll begin in 1954, with a fictionalized account of my father's meeting with Richard Feynman, since it led to the letter to Einstein, the 1960 paper, and ultimately the 1994 experiments. The meeting, it's safe to say, was a fact, as he told the story many times over the years without significant variation or embellishment. The details given here were regular parts of the story. He was always respectful of Feynman, if not awed, and he dedicated his final paper on the UFT to him (see p. 227).

PASADENA, CALIFORNIA, CAMPUS OF CALTECH

"Maybe," thought Richard Feynman, "*maybe* I should reconsider my open-door policy during office hours." Now here was a complete stranger sitting across the desk from him, a nobody who wanted to talk about a unified field theory—his *own* unified field theory, not Albert Einstein's.

This was going to be a waste of time, and Dr. Feynman, Professor of Theoretical Physics, didn't appreciate anyone wasting his time, unless it was himself. He continued bouncing a tennis ball against the wall of his office, barely glancing at the interloper, a dark-haired young man who seemed like a lot of other graduate students: far too serious. *Well, perhaps we'll play with him for a minute or two and send him on his merry way.*

"What did you say your name was?"

"Frederick Alzofon. I'm a great admirer of your work, Dr. Feynman."

"And where did you say you were from?"

"UC Berkeley."

"Working on your doctorate?"

"Yes."

"And what do we have here? Something about a unified field theory?"

"It's a draft of a paper."

"Well, let's have a look." Feynman caught the tennis ball in one hand and pulled the paper across the desk with the other. "Mind you, I can't read this in detail," he said as he scanned the pages. *Hm*, the guy knew his math, but he had an irritating way of numbering his equations: "*1.10, 1.25, 1.35...*"

"A dollar-ten, a dollar-twenty-five, a dollar-thirty-five..." Feynman counted off with mock seriousness.[10] He glanced up. The young doctoral student wasn't smiling. Maybe he was just intimidated. Feynman kept reading, flipping pages. Now he began to catch the drift of the fellow's thought. The little upstart wanted to start from scratch, sidestepping the general theory of relativity entirely and pulling a unified field theory out of *special* relativity!

Professor Feynman stood up, smacked the paper with the back of his hand and shouted, "You're a real *sonofabitch*, aren't you!"

"*You're* a sonofabitch," said the interloper without missing a beat. Weirdly unruffled. Not even angry. Well, this conversation had taken an odd turn, hadn't it?

"Where are you from?" Feynman asked.

"Santa Barbara."

"Before that."

"Detroit."

Ah well, that explains it, Feynman thought. He sat down again without another word and started reading. After a couple more pages, he glanced up.

"You eliminate the infinite zero-point energy."

"Yes."

He smiled. "Good. I like that. But where are you going with this? How wide is your net?"

"The theory includes classical and relativistic particle mechanics, electromagnetic fields, the classical gravitational field, a model of inertial and gravitational mass, the equivalence of matter and radiation, quantum mechanics, and a means of extending the theory without infinities to subatomic processes."

Feynman rubbed his chin. "Ambitious. But realistic?"

"The basic ideas aren't new, but they are well accepted. All I've done is put them together in a new way. And yes, I think it's realistic. It's *more* realistic than presently accepted theories, since the potentials of the fields characteristic of the theory do not possess any infinities, like the Coulomb

---

[10] Like most of the rest of the story, this part comes straight from my father.

60

potential and the static gravitational field—Newton's Law."

Feynman leaned back in his office chair and put his fingertips together in a steeple. "Well, you know a thousand times more than most people who come to see me. *But*—you've got a *long* way to go yet before this is publishable." He smiled.

"I know."

Always serious, wasn't he. Well, give him some encouragement.

"I think you might be onto something with this," said Feynman two hours later, after attacking the theory from a number of angles and failing to break it. "Physics needs a new and simple idea. Go home! Polish it up! Publish!" He smiled as he returned the young man's paper.

Frederick Alzofon shook Dr. Feynman's hand and left. The southern California sun was casting fuzzy shadows with its golden light as he walked across the campus toward his car. Students wandered between classes, and sat on benches and under shady trees, reading books. He envied them. He had wanted the academic life, the ivory tower, but the Darwinian reality on the inside, he found, was much different from the motto over the front gate, "Fiat lux" ("Let there be light"). Some professors—like Lenzen, Evans, and of course Feynman—stood for everything he held sacred. Others—like Oppenheimer, Lawrence, Teller, Birge—were enough to make him want out.

The talk with Feynman left him energized, his mind swirling with new ideas, new plans. "If Feynman didn't reject my ideas wholesale, maybe Einstein would be interested after all," he thought. "Well, why not? I'll write to him."

With a pregnant wife, two young children, and a new job as a physicist at the Santa Barbara Research Center, he barely had time to sleep as it was. Every spare moment was already taken up with his dissertation: *Multiple-Valued Functions and Sommerfeld's Method*. It still wasn't done, and it had *nothing* to do with unified field theory. But somehow he would find a way. He had to. Something like *this*, something that could make it easy to get into space? He could never let go of that. No matter what it took, he would do it. And aerospace was as good a place as any, if not better.

Chapter 11

1955 – A Letter to Dr. Einstein

THE FOLLOWING letter to Albert Einstein is dated February 22nd, 1955 (source: photocopy from A. Einstein Archive 59-093). It is evidently a clarification of an earlier letter sent in October, 1954, probably a few months after the meeting with Feynman, which my father said was in 1954. Unfortunately, the Einstein archive didn't have the October letter, which apparently included an early paper—perhaps the first—on the UFT. Einstein didn't reply, but this indicates nothing one way or the other, as he was by this time one of the most famous men in the world and busy with a flood of correspondence among other things. Unfortunately, Albert Einstein died April 18th, 1955, roughly seven weeks after the letter below was sent.

The October 1954 letter and accompanying paper, if they still existed, would probably give the flavor of the conversation between my dad and Professor Feynman described in *Chapter 10.* The 1960 paper on gravitation (Ref. 1a, p. 259) undoubtedly contains many of the same ideas. However, this was by no means their final form, as can be seen in *Chapter 37* (p. 227), and other papers referenced in *Appendix C* (p. 259, nos. 2, 4, 7).

SANTA BARBARA RESEARCH CENTER

SANTA BARBARA MUNICIPAL AIRPORT
GOLETA, CALIFORNIA
TELEPHONE SANTA BARBARA 85341

22 February 1955

Dr. Albert Einstein
Institute for Advanced Study
Princeton, New Jersey

Dear Dr. Einstein:

In connection with the paper entitled "Unified Field Theory and 'Virtual Processes'" sent to you 8 October 1954, there are some elucidating remarks which should be added to the manuscript. Unless otherwise stated, page and equation numbers refer to the ms in question.

A recent discussion with Prof. G. Y. Rainich at the University of Michigan has indicated that the term "quadratic simultaneity" is a misleading one. The motivation for introduction of the term was the existence of two equivalent derivations of the Lorentz transformation (cf. reference of note 9, "On the Electrodynamics of Moving Bodies," p. 46). That is, the concept of simultaneity leads to linear relations, among which is the fundamental one exploited in your paper (op. cit., p. 42)

$$t_B - t_A = \frac{x_B - x_A}{c - v}$$

and which lead to the Lorentz transformations (op. cit. pp. 44-48). On the other hand, the requirement that a spherical light wave have the same form and velocity of propagation in both a stationary and a moving reference system is stated in terms of a quadratic expression, and again yields the Lorentz transformation. My essential aim was to show that the two criteria were not equivalent if the light signal suffers fluctuations due to the additional processes named. More explicitly, the linear relation given above conceivably may be replaced by

$$t_B + t_B{}^0 - (t_A + t_A{}^0) = \frac{x_B + x_B{}^0 - (x_A + x_A{}^0)}{c - v}$$

and its companion relation (not given above), by

$$t_A' + t_A^{0\,\prime} - (t_B + t_B^0) = \frac{x_B + x_B^0 - (x_A + x_A^0)}{c + v^{\mu}}$$

these relations reducing to yours for a vacuum. Of course, a quadratic relation does not reduce to that of the special theory.

In summary, if

$$s^2 = r^2 - c^2 t^2 + r_0^2 \quad \ldots \ldots \quad (A)$$

denote the "distance" squared separating two point events, then if $s^2 = 0$, the events might be termed "quadratically coincident." I cannot find any term in the literature which conveys the meaning very well; in any case, some such alteration is suggested to replace the objectionable term in the ms.

Prof. Rainich expressed doubt as to the validity of eq. 1.12. However, I view the situation as analogous to that in which the quadratic form

$$s^2 = r^2 - c^2 t^2$$

is invariant under the transformation compounded of a rotation in space and the identity in time; the Poisson equation is invariant under such a transformation. Similarly, the quadratic form given above (A) is invariant under the Lorentz transformation in r and t, along with a rotation in $\{r_0\}$ space. The extended Poisson equation (eq. 1.33, p. 19) is invariant under rotations in both $\{r\}$ and $\{r_0\}$ spaces, with the identity in t and t_0.

The consequences of the discrepancy between observed and postulated states of motion are not sufficiently elaborated. Principally this was due to a reluctance to lengthen the paper beyond its present limits. It is my belief that the discrepancy results in the inertia characteristic of material systems, and, for that matter, gives significance to the quantity often referred to as the "mass" of a photon, $\hbar \omega / c^2$. Indeed, this appears to be an essential implication of Lorentz' discussion (footnote 10 of the ms), in its present application. A similar condition is met in the observation of an impedance characteristic of electrical networks: due to the existence of a degree of freedom in addition to those explicitly in evidence (input and output voltages), there is a discrepancy between input and output voltages. For example, a dissipative impedance is considered to arise from electron-ion collisions, and an inductive impedance from storage of energy in an induced field. With reference to light signals, the additional processes give rise to the need for additional coordinates since it is agreed that light signals establish the metric.

The reduction of

$$p^2 + p_0{}^2 = \left(\frac{E}{c}\right)^2$$

to

$$p^2 + (mc)^2 = \left(\frac{E}{c}\right)^2$$

for a free particle is considered to hold strictly in a rest system alone. The second relation would seem to hold as well for energies sufficiently small that creation and annihilation of photons, electrons, etc. has a negligible influence on the motion. Similarly for the general relation 4.9, p. 36.

The introduction of the de Broglie hypothesis (p. 17) and the mixture of a classical and quantum mechanical orientations was not discussed as fully as might be. Thus the extended Maxwell equations are introduced (pp. 23-26) and these are followed by the extended Dirac equations. I felt that this was not an inconsistency since second quantization can be applied to both equations, while the equations of motion of quantum mechanics merge into those of classical mechanics at the macroscopic level. However, the motivation for the introduction of quantum mechanics lies in the necessity of averaging over many possible system states, and from one point of view these arise from fluctuations in a given state of a system. Such fluctuations are an integral part of the formalism proposed, even on a classical level, without the need for further introduction of state averages. I then am led to wonder whether or not there is need for the quantum mechanical orientation. This last is purely speculation: it seems to me that the essential deductions of a preliminary study would follow, whatever the orientation. The intent of the physical model seems sufficiently clear.

The relation

$$\gamma_{k,\omega,x} = \gamma_{k,\omega,-x}$$

is properly a vacuum condition. It is, however, employed as a condition on the creation and annihilation of electrons and positrons since these processes are usually described in terms of a weak interaction, and hence a small perturbation of the vacuum state.[11] This is, of course, a proposal to replace the Dirac view of the nature of the positron with an alternate model.

It is noted that Bose-Einstein particles can contribute their effects to the relation 3.27. My interest at this point was in estimates alone and in obtaining Newton's law of gravitation, eq. 4.21.

With reference to Newton's law, the implication of the physical model is that the law holds only to a first approximation. Polarization of the virtual charges (of the effective gravitational charge) leads to gravitational mass renormalization. It is of interest to enquire as to the expected alteration of the orbit of the Earth's moon on this basis. The lunar orbit has been measured with great accuracy and it is known that Newton's law is not sufficient to account for observed anomalies.[12] It is my impression that the general theory of relativity predicts an effect of an order of magnitude insufficient to account for the discrepancies.

[*Ed note. The following paragraph, it should be emphasized, references an early version of the UFT, and it is doubtful my father continued to hold this view of gravitation, as it never appears again in his papers. As I recall from numerous conversations over the years, his ideas on gravitation became fully focused in the mid-1970s, when he was brainstorming the meaning of the 1957 sighting referred to in Jim McCampbell's book,* UFOLOGY, *(see p. 77). However, this is conjecture on my part.*]

A further consequence of the proposed theory is the alteration of a gravitational force near an intensely radioactive source. The change of sign of the inertial momentum for a positron (or for any anti-particle) is suggested to result in a reversal of the gravitational force for such particles in interaction with ordinary particles (e.g. electron, proton, etc.). I should like to raise the question as to the result of a positron screening of a nucleus in analogy to Coulomb screening of a nucleus by the electrons in inner orbits. If a radioactive nucleus which produces a space charge of positrons or anti-particle mesons could be found, would the gravitational force exerted on such material by the Earth be perceptibly altered? Again this is speculative to the point that I omitted any such proposal from the ms.

Prof. Rainich thought that there might be a possible equivalence of the nonlinear general theory of relativity, and the linear theory proposed. To date there has been no such comparison.

---

[11] W. Heitler, The Quantum Theory of Radiation (Oxford University Press, 1944) second edition, p. 193

[12] H. N. Russell, R. S. Dugan, and J. Q. Stewart, Astronomy (Ginn and Co., 1926), Vol. I, p. 289. Compare also various works by Prof. E. W. Brown (references are not immediately available to me for quotation)

I regret that the demands of other duties have prevented the further development of the theory in any significant manner. May I express my admiration for the clarity and physical insight displayed in your papers, which I have read with keen enjoyment. Much of the inspiration for the ms sent you has been derived from a desire to emulate the classic simplicity of your expositions.

Sincerely,

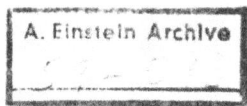

F. E. Alzofon

[A. Einstein Archive stamp, 59-093; jpegs of letter provided by Barbara Wolff, Einstein Information Officer, Albert Einstein Archives, The Hebrew University of Jerusalem. See *Acknowledgments*]

Chapter 12

1960 – The Air Force Takes an Interest

AFTER the meeting with Professor Feynman, my dad organized his entire life outside of work around one goal: writing a paper on gravitation. He chose to focus on gravitation because he thought that the theory might lead to a new propulsion technology that would open the door to cheap and easy space travel. This is what he told me in 1958 when I asked him what he was writing (I was eight years old at the time). In 1960, "The Origin of the Gravitational Field," was published in *Advances in the Astronautical Sciences.*[13]

I well remember the disciplined daily routine that prevailed throughout the interim between his meeting with Feynman and publication of the paper. It began the moment he arrived home from Lockheed, Sunnyvale. He would give my mom a kiss and make a beeline for the back of the house, where he had converted a stucco-lined, linoleum-floored storage room built by the former owners into a "fortress of solitude." It probably wasn't up to code—the ceiling was low and it was often cold and drafty—but he had covered the walls with soundproofing and bookshelves, and it was quiet.

Minutes after arriving home, he would lock himself in this inner sanctum and the typing would begin. The "do not disturb" warning was a standing order and meant to be taken seriously. The clacking of typewriter keys ended only for dinner, which he shared with my mom while the children watched TV in the next room with the door closed. After dinner was over, he would retreat to the study, close the door, and continue writing until eleven. Once a week, he broke with routine to sit down with the family and watch *Perry Mason.* "Perry Masonite" was cause for celebration, especially for me and my siblings, though we were repeatedly admonished to sit quietly throughout the show. After it was over, it was understood that he would lock himself in the study again. The study was sacred. A parakeet we acquired during this era knew the routine as well as any of us. The first phrase it learned was "Shut the door!" which it would squawk with glee, over and over, as it fluttered around the house. This amused everyone, including my father, but the after-work and dinner rules didn't soften.

Vacations, too, were just another opportunity for research. I remember my dad sitting on a beach chair by the campfire in Yosemite with a stack of physics books by his side, typing on a Royal portable, then rolling out the page and filling the gaps with equations written in jet black ink from a Parker fountain

---

[13] Vol. V, pp. 309-319, Plenum Press, New York, 1960

pen. In 1959, my mom gave me a recon assignment: I was supposed to ask my dad to write out his two most important equations. I found him sitting at his typewriter in the study. He was a little puzzled by my request, but complied. I remember him trying to explain the equations, which made no sense to me. It made no difference—I had gotten what my mom wanted. That weekend, she took me to a jewelry store where she got out the scrap of paper with the equations and explained to the jeweler that she wanted to engrave them on a pair of tie clips for his birthday.

The work rules never relaxed. They were in effect in 2000, when my dad arrived in Las Vegas for a vacation with a small library in tow. They were in force up until a month before his death in 2012.

The point of these anecdotes is not to elicit sympathy. My dad was distant, but my mom repeatedly told us that he was on an important mission, and we accepted his absence from everyday family affairs as normal. He wasn't averse to talking with us about his work on gravitation, either, even if relativity and the "creation and annihilation of charged particles on a subatomic plane" didn't make as much sense to me as the physics of two-wheeled bicycles, baseball, football, fishing, and dinosaurs.

"Gravity and antigravity," we understood even as children, were two words that summed up everything my father stood for as a scientist. About his work at Lockheed, we knew almost nothing. It was all classified and he never spoke about it. The atmosphere of Cold War secrecy made us feel as if the family was suspended on invisible threads, secure and insecure at the same time.

The reason I've painted a portrait of this period is simply to emphasize that his mission to conquer gravity was all-consuming, not only for him, but for the entire family, and more important, he was doing it all on his own. Lockheed had no idea what he was working on in his spare time and didn't care. Except for a couple of UC Extension courses on calculus he was teaching, his ties to academia had long since been dissolved. The few physicists he knew were not interested in unified field theory, especially *his* unified field theory. This is what made Feynman exceptional: He was a great scientist who had taken the trouble to listen, and had told him he might be on the track of something important, a simplification of physical theory (see p. 229). But the episode with Feynman was just that: an episode. There was no ongoing communication. The encouragement that Feynman gave, however, was enough to fuel an all-out effort that devoured every spare moment for decades.

In 1960, the theory of gravitation was published. Sixty percent of my life and one-hundred percent of my siblings' time on earth had been taken up in anticipation of that moment. We were all excited. When my dad showed me an article about the paper in the *Palo Alto Times*, a daily newspaper I delivered every day on a bicycle, I thought it meant that he would soon be emerging from his study to work on gravity and antigravity with other scientists. Surprisingly his paper drew attention, *not* from other scientists, but from the U.S. Air Force.

Were they inspired by a desire for cheap and easy access to space? Was it to recruit him for a secret program to back-engineer alien craft? None of the above. The motivation was Cold War paranoia.

It seems that in 1960, a Russian physicist named Kirill Stanyukovich was boasting that the Soviet Union had an "antigravity propulsion system" that allowed the cosmonauts to get into space without rockets and maneuver like fighter jets once they were out there. Naturally, this raised eyebrows at the Pentagon, where embarrassment over Sputnik in the 1950s had not been forgotten. They wanted to know if Stanyukovich was telling the truth, so they hired Garbell Research Associates of San Francisco to investigate his claims.

The lead investigator, Dr. Maurice A. Garbell, decided to study gravitation research, not only in the Soviet Union, but everywhere in the world since the time of Newton. My father's freshly minted 1960 paper appeared just in time to be included in the survey.[14]

Garbell concluded that the Soviets were sending up a trial balloon to provoke the U.S. into admitting that *we* had such a technology. Perhaps the Kremlin was responding to rumors of reverse-engineering of the disks recovered in the Roswell crash.[15] Garbell concluded that the USSR did not have antigravity, nor did anyone else, and they were not likely to get it, because the general theory of

relativity, which then, as now, was considered the gold-standard all over the world, offered no conceivable method for altering gravity.

One exception was noted: Alzofon's theory. The language in the report was encouraging but conservative. In a private conversation at a restaurant in Palo Alto (Stickney's, near Stanford stadium), however, Dr. Garbell said that the UFT was the only theory that he had looked at that "had a prayer of an engineering application." Garbell mentioned a possible avenue toward altering gravity in the report. My dad had already been thinking along the same lines,[16] but the favorable write-up offered him an excuse to approach the Pentagon. A meeting was arranged with the head of the Foreign Technology division at Ames Research Center just down the road from Lockheed at Moffett Field. My dad had worked at Ames in 1957 – 1958, at NACA, the precursor to NASA. Now he found himself sitting across the desk from the head of the Air Force's Foreign Technology Division, a full-

---

[14] "Soviet Research on Gravitation, an Analysis of Published Literature." AID Report 60-61, Sponsored by Science and Technology Section, Air Information Division, Distributed by U.S. Department of Commerce, Business and Defense Services Administration, Office of Technical Services, Washington, D.C., October 1960. Available online. While I have seen the report, I do not have a copy of it. As I recall, it was dry reading and contained no revelations.

[15] Most unimpeachable source: *The Day after Roswell*, by Col. Philip J. Corso and William Birnes, Pocket Books, 1997, ISBN: 0-671-03695-5

[16] Though the devil was in the details, Garbell's report did forecast the solution that appeared in the 1981 paper.

bird colonel he had never met before.

"Well, Dr. Alzofon, quite frankly, when you called, I figured you were some kind of nut case. But I did some checking, and the Pentagon says that if anyone could figure out how to do this, it would be you, so why don't you tell me what you have in mind." (This is close to a verbatim quote of what my father told me soon after the meeting and several times over the following years.)

My dad launched into his pitch. The concept—an antigravity propulsion system—was extraordinary, but he couched it in the conservative language he had learned in the laboratory at Cal, with a lot of deferential "ifs" and "maybes." But with the Air Force's own report backing him and the imperative to get out in front of the U.S.S.R. in space technology, not to mention the lingering embarrassment of Sputnik, he thought he had a good shot at launching a research program.

The Colonel listened and then he said, "Research is all well and good, but what I want to know is, *can you make me something that flies?*"

"I have no idea," my father said. Clearly, he had been laboring under the illusion that research programs are set up to *investigate* the unknown. Based on what research revealed, engineering goals would be determined. The Colonel said that regrettably the Air Force didn't have the money to finance a fishing expedition. They could only justify investing in things that had a predetermined outcome and a delivery date, such as a project to build a better jet plane or a faster rocket. "Stanford's right down the road," he said. "Maybe they'd like to start something."

Unfortunately, as Dr. Garbell observed in the conversation at Stickney's, my dad's theory would never be welcome at Stanford or any other university where general relativity set the boundaries for discussion (*see Chapter 38,* p. 241). As Garbell put it, "A snowball would have a better chance in hell."

The Colonel remained interested, however. He said he wanted to adopt my father's program and keep it on the shelf until the money became available.

"Of course, a project of this sort would have to be classified higher than the Manhattan Project," he added. "You might not even qualify for the clearance necessary to work on it." The Colonel knew my dad already had a Top Secret clearance and had worked on many sensitive projects.

"Let me get this straight," my dad said, his blood pressure spiking. "You want to 'adopt' my idea for a program that you can't fund, and once you do that, I might not even be able to work on it?"

"That's right," the Colonel laughed.

"My *own theory?*"

"Well, I'll see if I can get you authorization, but I can't promise anything." It was a moment straight out of *Catch-22* or *Dr. Strangelove*.

My father told the Colonel he wasn't interested and walked out the door with the Colonel's laughter echoing in the hall behind him. I remember the day he came home from that meeting, visibly shaken by the idea that his theory and the applied technology, which he had worked on for twenty years and intended for the use in the space program, might disappear down a black hole deeper than the Manhattan Project. He had gone into the meeting with high hopes and emerged in dread of losing everything.

Nothing further came of the meeting, but it was a turning point in a couple of ways. First, the Colonel's threat made a deep impression on my dad and influenced him to go public with the technology as quickly as possible in 1981. Contradictorily, it conditioned the paranoia and secrecy that surrounding the 1994 experiment. Last but not least, it was the first in what was to become an almost unbroken chain of disappointments that would span the next fifty-two years.

But the Colonel's all-important question: "Can you make me something that flies?" lingered on. The answer was not easy to find. It came thirteen years later, when my dad was puzzling over data gathered from a UFO sighting that occurred in 1957.

Chapter 13

1973 – A Key Sighting

MY DAD continued to speculate about an applied technology based on his theory of gravitation after the meeting at Ames Research. Since UAP (Unidentified Aerial Phenomena) clearly manipulated gravity and inertia, he looked to them for clues, but the typical sighting offered little in the way of hard data.

In 1973, quite by accident, I discovered a book called *UFOLOGY—New Insights from Science and Common Sense* at Kepler's Books in Menlo Park. Kepler's was a location of some renown in the Bay Area as a center of the peace movement and parking-lot concerts by Joan Baez and the Grateful Dead. The author was James M. McCampbell, Northern California Chairman of the American Nuclear Society, a nuclear engineer who worked for Bechtel and NASA. McCampbell lived in Belmont, a short hop up the Peninsula from Menlo Park, and Kepler's Books was probably one of the few places in the world where his book was available.[17]

UFOLOGY was and *is* even now a landmark classic, the first book on UAP to take a rational, scientific approach to the phenomenon. McCampbell wasted no time with the "Are they real or not?" debate. In the introductory chapter, "Certified UFOs," he noted that decades of similar reports, radar echoes, photographs, films, and trace evidence from independent witnesses all over the globe made it safe to assume that UAP were material craft. Then he analyzed the physical evidence in chapters with titles such as "Vehicle Design," "Composition and Luminosity," "Sounds," "Electrical Interference," "Physiological Effects," "Flight and Propulsion," and so forth. This was the kind of analysis my father had wanted to do, but had never had the time.

My eye was immediately drawn to a chapter called "Microwave Propulsion." Among other things, it had some remarkable observations stemming from a 1957 sighting aboard an Air Force B-47. This sounded like the kind of hard data my father had been craving, so I bought two copies of the book and forwarded one to him in Long Beach, where he was working for Rockwell International.

---

[17] The first edition was self-published by Jaymac Company, Belmont. It was later picked up by another publisher and sold as a mass-market paperback. The entire text can now be found online.

The incident occurred on a routine B-47 training flight over the Gulf of Mexico. The B-47 is familiar from Stanley Kubrick's classic, *Dr. Strangelove*. But delivering nuclear weapons was not its only mission. This particular flight was an electronic countermeasures training operation.

The pilot radioed in to report that a UAP "as big as a barn" with a "steady, red glow," was flying rings around the six-engine, turbojet-powered B-47, which was moving at speeds greater than 500 miles per hour (the top speed of a Boeing B-47 Stratojet is 608 mph). The sighting lasted more than one and a half hours while the B-47 flew across Mississippi, Louisiana, and northern Texas.

The B-47 had electronic monitoring equipment onboard, including an ALA-6DF passive receiver with back-to-back antennas spinning at a rate of 150 – 300 rpms in a housing on its belly. Incoming signals were displayed on an APR-9 radar receiver and fed into an ALA-5 pulse analyzer. The equipment showed that the UAP was emitting powerful bursts of microwave radiation in a very narrow range. The following summary was given by the Wing Intelligence Officer at Forbes Air Force base:

- Frequency: 2995 to 3000 Megacycles per second

- Pulse width: 2.0 microseconds

- Pulse repetition frequency: 600 cycles per second

- Sweep rate: 4 rpm

- Polarity: vertical

These figures are important, because they showed up again at the end of a series of equations my father used in preparing the 1981 paper and again in designing the 1994 experiment. In other words, he arrived at identical parameters for gravity-control technology by an independent route based solely on his unified field theory. This startling coincidence was the first of many. The correspondences between observations of UAP and predictions of the UFT form a network of mutually reinforcing evidence, somewhat like a long, complex password to unlocking the secrets of alien technology.

McCampbell goes on to say, "The flood of microwave energy from the UFO was an essential, integral part of a propulsion system that is common to all UFOs."

But back to the year 1973: Here, in tantalizing form, was a major clue to what my father had been seeking since 1960. Microwaves played a role, but what was it exactly?

My father said that while he found the data intriguing, he wasn't sure what it all meant. Eventually I made contact with Jim McCampbell and introduced him to my father. The friendship lasted until Jim's death in 2008 (see *Chapter 14,* p. 84, *and Chapter 33,* p. 205). My dad agreed with McCampbell that microwaves were an essential part of the propulsion system, but it took another lucky break to fill in the missing piece of the puzzle and trigger that "Eureka" moment.

The second break occurred in the late 1970s. Prompted by his conversations with McCampbell and ceaseless contemplation of the data revealed in the 1957 Gulf of Mexico sighting, he was browsing the shelves of a technical library, "looking for a certain state of matter," as he put it, when he stumbled across *Dynamic Nuclear Orientation*, a textbook by C. D. Jeffries (Interscience, John Wiley & Sons. N.Y. 1963). Something about the title immediately grabbed his attention and he took it down from the shelf. As soon as he fanned it open, he knew he had the answer he had been seeking.

"In principle," my dad wrote, "the method of dynamic nuclear orientation is easy to state. A constant magnetic field is imposed on a specimen of a ferromagnetic material, causing the electrons of the atoms to precess about the direction of the field with a characteristic (Larmor) frequency. An oscillating magnetic field which varies at the Larmor frequency is then applied to the specimen at right angles, causing the electrons to tip over and become oriented. To preserve the angular momentum of the specimen, the nuclei must also tip over and become oriented."

The "oscillating magnetic field" was supplied by square-wave pulses of microwave radiation in the vicinity of 3000 Mhz.

Suddenly the rationale for the propulsion system fell into place. The description is detailed in *Chapter 17*, p. 99, and many other places, but in broad terms, rapid cycling of nuclear orientation acts like a pump to draw energy out of the gravitational force in the vicinity of the vehicle. A similar technique is used in cryogenic cooling (adiabatic demagnetization of paramagnetic salts). The difference, of course, is that the cryogenic process operates on a molecular level, but "gravity cooling" operates on a subatomic level against the physical source of the gravitational field.

One difficulty presented itself immediately: dynamic nuclear orientation as described in the text required cooling the specimen to the temperature of liquid helium. Obviously, UAP were operating at higher temperatures than that, though there is evidence that they periodically take steps to supercool the hull, or they are forced to land when heating disrupts the gravity-control effect. However, he saw a way to overcome this problem using a "very pure isotope of aluminum," which has a slow thermal decay time for nuclear orientation. Colloidal iron or chromium is added to the mix. The nuclei of these metals are easily oriented, but have a rapid thermal decay time.

The orientation of the iron or chromium molecules transfers to the aluminum by conservation of angular momentum.[18] This, too, was suggested by data from UAP: Physical evidence had already been found to suggest that aluminum and colloidal iron particles are principle components in the hulls of the flying discs.

With Jeffries' book in hand, he ran some calculations. When he was finished, he found that his figures matched the data gathered by the B-47 over the Gulf of Mexico. This had to be more than coincidence.

The key had been turned in the lock, the tumblers had clicked, and a door had opened to a future beyond imagination.

---

[18] How does angular momentum transfer in a system with negligible mass, that is, a system under the influence of gravity control? In *Ch. 19*, p. 119 - 120, *Ch. 21*, p. 135, *Ch. 31*, p. 197, *Ch. 34*, p. 216, *Ch. 35*, p. 220, *App. B*, p. 256, and in published works listed in *Appendix C* on p. 259, my father makes it clear that the agency is the magnetic moment: "We can imagine a procedure for lowering the energy density of the gravitational field (and therefore reducing its strength) by making use of magnetic moments."

Chapter 14

1981 – Tesla Country

AS THE 17ᵗʰ Joint Propulsion Conference neared, my dad rushed to put the finishing touches on a massive paper on gravity control, rolling up his sleeves after a long day at Boeing and working evenings and weekends deep into the night. Nothing about his after-work routine had changed since the 1950s and 1960s, but now he lived in Seattle. In July, he scheduled a few vacation days, bought a plane ticket and paid for a hotel room in Colorado Springs with his own funds.

More than just a rationale for the device and the method of instrumentation, the paper included a weighty theoretical foundation aimed at justifying the UFT as an alternative to the general theory of relativity. This part of the paper was tailored for an audience of physics professors rather than the aerospace engineers and scientists who would be gathered at Colorado Springs. But, as is somehow common in publishing, a conspicuous blemish made its way onto the cover, in the form of the word "anti-gravity," with a spelling borrowed from the British lexicon.

Years later, my father would say, "I didn't know it at the time, but one mention of 'antigravity' and you're sunk. Their eyebrows shoot up, they get a bemused grin on their face and that's the end of it." What ended, of course, was his claim to credibility. And, since the 1981 paper was to become his calling card in scientific circles for the rest of his life, the presence of "anti-gravity" in the title became a significant impediment to progress.

Prior to the Joint Propulsion Conference, my dad had retained the services of a patent attorney. By the time of the conference, the patent had been submitted and the term "pat. pending" appeared in the paper. This put his mind at ease about describing the technology in public. Nothing was left out: the paper included a modular circuit diagram and a complete theoretical foundation that would enable physicists and engineers to imagine new configurations for the propulsion system and calculate the specifications for a vehicle. None of this bothered my father in the least. Indeed, his whole strategy was to go public with gravity-control technology first, set off a stampede, and enjoy the show as humanity poured through the open door into space. He hoped that he would be able to head up an R&D program and hitch a ride on one of those spacecraft, too.

But humanity—or shall we say, the members of society empowered to do something—didn't share his vision.

Initially, Boeing had requested that he put the company name on the cover of the paper. At the last minute, they withdrew their approval. But copies of the paper had already been printed and bound, so he spent the evening of his arrival sitting on the bed of his hotel room lining out "Boeing Aerospace" with a felt-tip marker on all the copies he had brought along with him. It was a somewhat comical gesture, since "Boeing" remained clearly visible.

Years later, there were rumors among ufologists that Boeing was working on "gravitic[19] propulsion," and I always wondered whether my dad's paper was the source of that rumor. If so, it was bitterly ironic, because Boeing wanted nothing to do with the 1981 paper, and *doubly* so because an investment of corporate pocket change would have put them at the forefront of gravity-control development and made the company untold billions of dollars—*in my humble opinion*, naturally.

The following day was the day of the presentation. He discovered that he was last on the program. By the time he stepped up to the podium—at 9 PM, as I recall—the hall was almost empty. Two hecklers sat side-by-side in the second row and began interrupting him, smirking and laughing during his opening statement. He invited them to explain what they found funny, and when they shrugged like schoolboys, he asked them if they knew anything about the Lamb-Retherford shift, and when they failed with some embarrassment, he asked whether they could define a light signal. Silence. They sat still and listened from then on. He always wondered whether or not they were CIA plants, since drunken hecklers would normally have left after being humiliated, but they sat through the entire talk.

I had always been optimistic that the Joint Propulsion Conference would be a turning point. After all, Colorado Springs was the home of Tesla's legendary laboratory, an auspicious launch pad for a new technology, and the paper offered a complete recipe for a revolutionary new propulsion system that would put us—the U.S.—light years in front of any nation on earth in the exploration of space. How could NASA or the military-industrial complex resist? My father had an impeccable record of problem-solving for Lockheed, NASA, and Boeing, and he was still working in the aerospace industry. The 1960 meeting with the head of the Foreign Technology Division at Ames would have gone on his record, along with the Colonel's favorable judgment. All it would have taken was an interdepartmental phone call, a summons from Washington, a nod from his employer.

But the call never came. Nor did the 1981 paper provoke any curiosity in the scientific community at home or abroad, where my father was more optimistic it would be well-received. Instead, the silence was deafening. There was no criticism, no compliments, not even a cough in the empty room. That's not quite true: A couple of letters filtered in from graduate students in foreign countries who asked

---

[19] Surely one of the most egregious attempts at word coinage in the history of the English language.

questions that revealed how little they had understood its message. The same thing happened in 2003 when a Silicon Valley friend, the late Richard Karpinski, made an effort to stir up some discussion of the UFT in physics forums. The comments were so ill-informed that my father refused to answer them. I had always thought that his message was clear enough, but as I read the comments I saw that the writers were anxious to superimpose their own knowledge of general relativity and quantum theory onto the UFT, at the expense of the latter. If they had expended half as much energy simply *reading* the text of the 1981 paper, or *The Unity of Nature*, his 1993 magnum opus on the UFT, they would have had all the answers they wanted and more.

For a while, my father rested and regrouped. Then he began writing and sending out proposals to nonprofit organizations. As noted above, he began to realize his error in using the word "antigravity" in the title and began using "gravity control" instead. Jim McCampbell was a friend by this time. As a former chairman of the American Nuclear Society in northern California, Jim suggested sending a proposal to the Department of Energy. He also offered to write a cover letter. We will close this chapter with the text of his letter, which was addressed to W. Kenneth Davis, Deputy Secretary, Department of Energy.

According to Wikipedia, Davis (1918 – 2005) was an American chemist, a leader of the World Energy Council, former vice president of the National Academy of Engineering, former U.S. Deputy Secretary of Energy, and director of reactor development in the Atomic Energy Commission. He was elected to the National Academy of Engineering in 1970 "for contributions to the development of nuclear power technology and its industrial application." There is no indication of his response (if any). However, McCampbell's confidence led my father to write the research proposal and submit it to the DoE. I was unable to find a copy of the proposal among his papers, but there is no doubt that it was sent and returned after eliciting the usual lack of interest. The proposal was probably similar to *Chapter 35*.

It was, however, *copied* by the DoE. My dad put a microdot on the back of one of the pages and the microdot was gone on the page he received back. In other words, one or more photocopies had been made, and they had returned one of the photocopied pages to him by mistake. What it all means is anyone's guess.

In 1981, I saw Colorado Springs as a propitious location to announce the discovery of gravity control because it was the site of Nikola Tesla's renowned Experimental Station. In 1899, Tesla had said, "Progress in this field has given me fresh hope that I shall see the fulfillment of one of my fondest dreams; namely, the transmission of power from station to station without the employment of any connecting wires."

If I had looked deeper into Tesla's career *after* he left Colorado Springs in 1904, I might have been less sanguine about the future of the 1981 paper: Tesla packed up and left because he had run out of financing, and in spite of some astounding demonstrations, he never achieved his dream of worldwide wireless power transmission. Regrettably, his greatest invention was buried with him, as no-one has been able to replicate what he accomplished at the Experimental Station. As fate would have it, my father left Colorado Springs without making a splash, and in spite of an astounding demonstration of the technology in 1994, he, like Tesla, never lived to see the realization of his dream.

This book is predicated on the hope of ending the unfortunate parallels between my father's work and Tesla's. One hopes that we will soon see gravity control spreading around the globe like a wildfire, powering a new renaissance in the environment, the economy, space exploration, and expanding the frontiers of human consciousness.

A LETTER TO THE DoE

A scan of Jim McCampbell's letter appears on the following page. The letterhead was redacted to remove address and telephone number.

JAMES M. McCAMPBELL'S 1982 LETTER TO THE DoE

McCAMPBELL MARKETING COMPANY
Products for the home.

November 19, 1982

Mr. W. Kenneth Davis
Deputy Secretary
Department of Energy
Washington, D.C. 20585

Dear Ken,

Here is something that I feel justified in calling to your attention.
A friend (Frederick E. Alzofon, MS Physics and PhD Math UC Berkeley) has
developed a concept for influencing gravity. He proposes to experiment
with currently available equipment that he will modify. His application
for a patent reportedly has no interference. Enclosed is a copy of a paper
on his approach. It is unlikely that renowned scientists will recognize
its value as it departs from tradition.

The reason that I think he may be on the right track is:

a) From theory alone, he derives the electromagnetic environment
   in which changes in gravity and inertia are predicted, and

b) Fifteen years of studying field observations of UFOs involving
   anti-gravitational and anti-inertial performance has independently
   led me to describe the same conditions.

Our government may have deciphered and duplicated the technology of
UFOs. If so, the above is of no consequence. If not, Fred's work could
be of critical significance. You can easily appreciate the range of implications. We have prepared a research plan requiring about a year and about
$300,000. The next step is to obtain financial support. Meanwhile, Fred
is proceeding on a poor-boy effort that is promising.

I think that this matter deserves serious attention and wonder if DoE
might have some interest. If deemed worthwhile, Fred and I would be pleased
to meet with you or Department scientists in Washington or on the West Coast.
We would waive any fees at this time but would request reimbursement of
expenses.

With utmost respect,

Jim

James M. McCampbell

*This should not be
passed around.*

Chapter 15

1994 – The Experiment

TEN MORE years were lost in sending out proposals and contacting investors. A pattern emerged: polite interest would be expressed, talks would follow, sometimes meetings. The prospective backers were always impressed. But when it came time to put money on the table, they would vanish, often saying, "Prove it out and get back to us." The change in attitude often seemed to occur after the investor consulted with an expert in general relativity.

It was intensely frustrating to be working in Silicon Valley during this period, a world awash with investment capital, where endless talk of the future and "visionary technology" was heard in the coffee shops and brass-rail barrooms of Palo Alto, where investors risked tens of millions on bonehead ideas with "sizzle," but no-one could be found with the courage to back a revolutionary new technology with solid roots in known science and the potential to outstrip Microsoft or Apple in terms of profit, and to mention its impact on culture, the economy, and the environment.

The reasons were many. As mentioned elsewhere, the term "antigravity" provoked suspicion. My father's lack of university connections played a part, as did the dismissiveness of "experts" who were consulted for a second opinion. Investors never seemed to be aware of the contradictions built into their pet phrase "Prove it out and get back to us." Why would we need them at all if we were able to "prove it out"? Would we need them to swoop in and take the lion's share of the profit perhaps? That seemed to be all they offered. The reality was that once my dad had "proved it out," they would have been doomed to stand in line with hat in hand. But venture capitalists, as I learned, are not daredevil gamblers with bags of loot. Their capital is reserved for ventures with minimal risk and quick rewards.

Somehow, in the midst of this exasperating, repetitive exercise in futility, my father found a couple of allies who were not wealthy but were keen to take action. They were, in fact, much like the ideal audience envisioned for this book. One of them had a friend at a major university who could give them access to a basement laboratory and most of the necessary equipment, especially the most expensive component: an electron paramagnetic resonance device used in assaying organic molecules. The missing components were procured for a modest price through electronics and surplus stores.

Once the decision had been made to go to work, the experiment came together with astonishing rapidity, which was curiously refreshing, after ten years of dithering by investors whose courage resembled that of a field mouse.

So finally, at long last, my father was doing what should have been done long ago, assembling the apparatus and advancing toward a denouement of sorts. As for me, I would know, one way or the other, whether or not his science-fiction dreams held any promise. As the date of the first experiment neared, I volunteered to fly up and videotape the whole thing, as it might turn out to be "an historic event," as I told him. My father, however, wouldn't allow it. He gave me a number of excuses, but I think it came down to his fear that the government would raid the lab, arrest everyone, and shut him down. His fears were not entirely groundless, but as it turned out, nothing happened, and the lack of photographic documentation turned out to be a grievous problem in the future, this book included.

Three rounds of experiments were conducted, on May 26th, June 17th, and June 18th. Afterward, my dad would say nothing until I signed a nondisclosure agreement. Once I signed the NDA, he said, "It worked the moment we threw the switch." He had little else to add until I came up to Oregon on a visit. Even then, I didn't see the experimental report contained in this book until 2001, the date of the postmark on the envelope he sent during my final campaign to find investors in Silicon Valley.

Though it was tempting to give the secret away to anyone who would listen, I kept quiet about the experiment for twenty years, including two years after his death. Then a friend who was a patent attorney told me that gravity control had become prior art in 1981, with the publication of the AIAA paper. In addition, my dad had published the secrets of gravity control several other places in the intervening time (see *Appendix C*, p. 259), and acknowledged that the experiments had been a success. Finally, the climate crisis and an overwhelming sense of frustration led to the decision to publish this book—against my better judgment. But when this invention might be the last hope of a doomed planet, what was there to fear, anyway? Then again, perhaps I've been misled by blind faith in my father's track record and a set of computer readouts. False hope isn't unusual or unheard of, but in this case there's reason to believe that these hopes are grounded in reality.

The report that begins on page 135 details the results of the experiment. The rest of this chapter will describe the unfortunate aftermath.

Within a few days after the experiments had been run, the Chemistry Department discovered that some expensive pieces of equipment had migrated from the laboratory on the second floor down to the basement, and they requested their immediate return. A decision was made among the conspirators not to inform the chair of the Chemistry Department that an earthshaking experiment had taken place in their basement. Instead, they decided to seek investors and reap the rewards.

There was only one small problem: *Now there was no working device*. None of them had the $200,000 it would take to replace the EPR module, let alone the remainder of the borrowed equipment, so that meant assembling everything from scratch, which was too expensive without an investor, which now had to be someone willing to buy in *without* a demonstration.

Meanwhile, the money squabbles began. One member of the trio declared that under no circumstances would he ever work with the third member of the group again. The question of how to split the profits from any future ventures then became enormously complicated.

Incidentally, I am not being coy about the names of the participants. My father mentioned them to me a few times in 1994, but after the group fell apart their names seemed unimportant, so I forgot them. They were not mentioned again, and when I received the written report in 2001, my father had whited out their names before making photocopies. Like the apparatus, they evaporated into thin air.

It had taken forty years for my dad to roll his Sisyphean boulder to the top of the hill, and now it had rolled all the way back down again. His refusal to allow me to film the experiments and the absence of a working device meant that there was virtually nothing to show to potential investors other than a handful of poor-quality photographs of antiquated equipment and the computer readouts. When I told him I wanted to keep looking for backing in Silicon Valley, he said okay, but I could not discuss the results of the experiment with anyone unless they signed an NDA. As I soon discovered, *no-one* would sign an NDA unless for a major corporation, so I was caught in a Catch-22.

For a time, Silicon Valley inspired me like nothing before or since. It was a fiery cauldron of new ideas and noble dreams, *all of them based on electronics*.[20] And here, in my hot little hands, was an electronic device straight out of a geeky sci-fi freak's fever dream. It was the key to a *Star Wars* or a *Star Trek* future, capable of doing for transportation and space exploration what the personal computer had done for commerce and communication, and it had a Silicon Valley pedigree. By a strange coincidence, I knew two Silicon Valley visionaries capable of appreciating that analogy *and* turning the dream into a reality. Everything about it seemed foreordained. And then, at the very doorstep of the future, they both failed to get it, or perhaps I lacked the salesmanship to get through to them. Whatever the reason, they looked at the science with suspicion and, like the peer reviewers at the big-name journals, refused to listen further.

It will always seem a great tragedy to me that gravity control did not launch in Silicon Valley, its place of birth and natural home.

Over and over again in the quest for backing in the years 2000 to 2007, the lack of a working device arose as an insurmountable obstacle. That my father had it in hand in 1994 counted for nothing, because, like a UFO, it had floated off into the misty realm of "witness testimony," leaving aught

---

[20] See *What the Dormouse Said*, by John Markoff, Penguin, 2005.

behind but the report you see on page 135. Even Hollywood, which was the last area I investigated before writing the book, insisted on "seeing something float" at the end of the show, which seemed like an interesting standard for a business dedicated to illusion, including *unreal* reality shows, CGI, science-fiction, flying saucers, and chasing UFOs (but never catching them). I must have asked myself ten thousand times what magic words would have broken through, not only to my former bosses Jef and Steve, but to all of these potential backers. In a sense, this entire book is a last-ditch effort to find those words before it's too late.

The need for a working model, plus the unwillingness of investors to spend anything on its creation remained an unbreakable deadlock from 1994 onward. By 2005 my father said he was too old to lead a project and he wanted to dedicate his remaining time to writing papers on other topics. I pleaded with him to allow me to keep searching, and he agreed as long as it was understood he would not leave home to guide the project. I widened the search to embrace show business and private space exploration companies, all the while seething because I knew that any average university chemistry lab had all the necessary equipment sitting on the shelf and all the brainpower on hand to assemble it. The whole thing could have been done in six weeks, if anyone had a mind to do it.

In the end, it was all to no avail. Meanwhile, just as in *The Wizard of Oz*, the sands of the hourglass were running out.

Chapter 16

2012 – Endgame

"I GUESS there isn't much to say."

These were the only words my dad could muster when I arrived at his hospital room in Corvallis in early December. It had been only two months since I had seen him, but the change in his appearance was shocking. What struck me most, however, were his words, or the lack of them. Though he was 93, he had always been fully lucid and ready to hold forth on all manner of subjects. Until just before the end, he had been working on an expansion of Sommerfeld's Method. And now he had only seven short words. Until I had spoken to his doctors and seen the X-rays of the tumor growing in his lung, I had held out hope that he *could* and *would* recover to continue his work. It was inconceivable that his body would quit before he had succeeded in repeating the 1994 experiment. But now I knew beyond a shadow of a doubt that his lifelong quest was coming to an end—an *unjust* end.

It will be difficult for the reader to grasp how my dad's all-consuming passion to unlock the secret of gravity and "antigravity" shaped the consciousness of our family. His quest may have been scientific, but it was comparable to a religion in the faith it inspired, a faith that my mother instilled in us from childhood on. The entire family believed that one day he would attain his goal and open the space frontier, and that this goal justified whatever sacrifices we might make. *Nothing* mattered more. The constant shadow of this quest had somehow made clocks stand still. Time could not advance until his theory had been proven or disproven. But of course time was advancing, whether or not I accepted it.

I had witnessed many ups and downs with the UFT and the applied technology over the decades since 1955, but I had never lost faith that in the end he would prevail. He was too good a scientist, he had solved every scientific problem he had ever tackled, and the 1994 experiment strongly suggested that his theory was more than a blackboard dream.

The pattern of his research was well established. First came the riddle, then the theory, then the doubters, then vindication, over and over. It had gotten so familiar that whenever the world said he was wrong about something, I took it as a reliable indication that he would soon be proven right. For example, the pattern had held when his experiments with infrared in the 1960s had vindicated his theory about the viability of thermography. It held when he had shown that the Sommerfeld method was extendable to objects of arbitrary shape. It held when he used optical methods to describe heat conduction in solids. And it held when he had described the physical origins of turbulence in the language of mathematics.

But now, at the end of his life's journey, he had nothing to say about his central quest, which, like the others had been successful, but *unlike* the others had gone completely unrecognized. Nor did he have anything to say about what must have been his monumental sense of frustration with academia, government, and private industry. He had nothing to say about what *might have been* if anyone—any department chair, any bureaucrat, any "visionary" billionaire—*anyone at all*—had understood the implications of the theory or the technology and had done something about it.

The planet might have been saved. We might have joked about it once in a while, but it was always tacitly understood that these were the stakes. Nothing more, nothing less.

And what now? No final advice about what to do about his technology once he was gone. He just wanted to go home and rest. "Forever," as he said.

I had taken a flight up to Portland, but after he died, I decided not to fly back. Instead I bought a train ticket at Albany, about twenty miles north of Corvallis, and rode to Los Angeles. It would give me twenty-nine hours of clacking railroad tracks, jostling passenger cars, and fitful sleep in a resolutely upright seat to absorb the catastrophic sense of loss and decide what must be done. Because there was no doubt in my mind that it was not over.

By the time I got out at Union Station in Los Angeles, I had concluded that it would be best to continue to do things *his* way, but with one small change: I would lift the veil of secrecy on the 1994 experiments.

As 2013 began, I started writing a proposal that included a description of the 1994 experiment, concentrating on the results, but omitting the nuts-and-bolts details. What physics department could resist the temptation of an easy Nobel Prize and a place in science history using nothing more than off-the-shelf equipment? In short order, we would be standing the world on its head.

Or so I thought.

Writing the proposal took seven months, and it was a waste of time. Not one of the six or so universities I wrote to, including my father's alma mater (as well as mine), the University of California, would consider it. They didn't "reject" it—rather, they refused to acknowledge it. It was just as Dr. Garbell had predicted in 1960: A snowball had a better chance in hell than the UFT had in a university. As for the idea of altering gravity, one did not even *suggest* such a thing at that level. No reason was ever given, nor was there any curiosity over the experiment. A non-Einsteinian view, let alone "antigravity," was simply beneath consideration. Past experience led me to believe that the source of the proposal (a nonacademic) alone was enough to preclude consideration.

Meanwhile, every day brought more news indicating that the pace of climate change was accelerating, which lent an air of desperation to the search.

When I realized that I would never get anywhere knocking on doors at the university, I decided to revisit private industry. I contacted, or attempted to contact, all of the companies engaged in privately funded space exploration, several of which were in my backyard in southern California. It was exactly the same: no response—not favorable, not unfavorable, but *no* response at all.

It was astonishing to find that lifting the veil of secrecy, something my father had feared would generate a cold-fusion-like fiasco, had no effect whatsoever. One professor—a talkative chap compared to his peers—was kind enough to send back the proposal and assure me that the university would do no work on my father's technology. I was grateful to know that there was, contrary to all other indications, a human being out there, but I could almost feel the touch of kid gloves reserved for the lunatic fringe. Beyond frustrating, it was terribly sad.

Writing letters, sending proposals, and waiting for responses had been enormously time-consuming. After three years, I was running out of options. In the second half of 2015, a new idea came to me: *Hollywood.* If everyone thought my father's invention was fantasy or science-fiction, fine—I would pitch it to people who loved fantasy and science-fiction. At least their publication standards would be less stringent than a scientific journal.

In 2005, I had stopped in at a magic shop in San Francisco and had an inspiring conversation with the owner, who encouraged me to contact big-name magicians with the idea of allowing them to perform *real* magic onstage. After all, what if a magician could toss an SUV like a football, or make members of the audience float, with no strings or wires attached? After two years, nothing had come of these efforts because direct talks were impossible to arrange. The people I needed to reach were as remote as royalty, surrounded by walls and fences and flappers whose whole mission in life was to prevent people like me from talking to them directly. Since the subject matter required some discretion and background explanation, it was far too difficult to penetrate the castle wall. The whole experience left a bad taste in my mouth and that had kept Hollywood far from my mind.

One night in 2015, however, I was watching a late-night program about UFOs, and as I listened to a couple of well-meaning scientists awkwardly attempt to interpret UFO flight with the limited tools at their disposal, a light switched on. Perhaps instead of *chasing* UFOs, Hollywood would like to *catch* one? If I could get just get one cable-TV documentary on the air, I thought, controversy would begin to swirl, and I would follow up with a book such as the one you're holding, that is, with a complete guide to building a gravity-control vehicle. It shouldn't be too difficult, should it? After all, if the Kardashians could generate millions of dollars, surely it would be a snap to sell something as awesome as a solution to the UFO enigma and a plan to save the planet. Much of the content envisioned appears in *Part II – History of an Idea* (p. 57) and *The Top-Ten UFO Riddles* book.

The Hollywood campaign continued through the rest of 2015. The first company I contacted had produced several made-for-cable documentaries about UFOs. I received the modern corporate response, which is to say, *no* response at all—just like the universities and the space exploration companies. Roughly ten other carefully vetted contacts followed. My only form of contact was formulaic rejections by email. I was ready to give up when, much to my surprise, I received an enthusiastic response: "We want to do your show." I replied by thanking them and suggesting we meet and discuss—and never heard from them again. No reason given.

It was at that point that I gave up on Hollywood, not because I was discouraged, but because I was damned sick of hemorrhaging time. Hollywood is home to some of the smartest people on earth, but the subject matter of *Gravity Control* isn't something that easily reduces to a log line and a ninety-second pitch, even if it does feature perennial favorites, such as flying saucers, aliens, a maverick scientific genius, and a plan to save the planet. If they didn't get *that*, what was the point?

Now I was faced with the grim fact that four years had gone by since 2012—four years of knocking on doors, making calls, writing letters and proposals, and sending packages. I was *out of time*, and so was the world, with its bovine indifference, blind ignorance, and institutional paralysis. Fortunately, there was a way out. Print-on-demand technology made it possible to shift into four-wheel drive, pull around the gatekeepers, and ride by in a cloud of dust with middle finger held on high. Just thinking about it gave me the refreshing sensation of having the wind at my back for a change.

A self-published book was the only alternative that made sense. A book proposal would have taken six months to write, above and beyond the book itself, and then it would hit the desk of a New York publisher, where, if it wasn't dismissed immediately for one reason or another, it would have been referred to an "expert," with the same tiresome result (see p. 241). The manuscript might have been *approved*, of course, but the low probability of success, coupled with the time loss and the accompanying risks, was simply too much to bear. Print-on-demand guaranteed publication, at least.

Once the book was an accomplished fact, even on the modest scale a self-published work, perhaps it would draw enough attention to trigger alarms in the halls of the institutions that had ignored all the formal, dignified entreaties over the last fifty years.

But here's the cool thing: If the book had no effect whatsoever on those hallowed institutions, there were still plenty of people *outside* the castle walls with the knowhow to repeat the 1994 experiment. At the risk of repeating myself *ad nauseum*, any E.E. with the knowledge required to design a microwave oven could do it.

BOOK II

Theoretical Foundation and Applied Technology

By Dr. Frederick Alzofon, Commentary by David Alzofon

You never change things by fighting the existing reality. To change something, build a new model that makes the existing model obsolete.

— Buckminster Fuller

Part III

From Gravity to Gravity Control

Chapter 17

Gravity Made Simple

ROBERT HEINLEIN famously said, "Never worry about theory as long as the machinery does what it's supposed to do."[21]

Unfortunately for the many who've tried and failed, the Heinlein approach is no way to build a gravity-control device. If one endeavors to control gravitation, *everything* is predicated on a firm understanding of the causes of gravitation, rooted in historical science—in other words, a good theory. Even that is not enough: The relationship between gravity and the other fundamental forces in the universe must be understood, in other words, one must have a *unified field theory*. Otherwise, how can one know what levers to pull or what results to expect?

Most of those who stalk the antigravity unicorn—even those who think they know the scientific method—tend to underestimate the importance and demands of theory. The subject is significant enough to warrant a separate chapter (see *Chapter 30*, p. 181). Here, however, such a discussion would only impede our progress. The aim of *this* chapter is to provide a clear and simple version of my father's theory of gravitation as an orientation point for the rest of the book. I call it the "paper-napkin version" of gravity and gravity control after a saying my dad repeated many times over the years: "If your concept isn't *visualizable*, if you can't explain it with some sketches on a paper napkin, some hand-waving, and a few commonsense analogies, then your idea is probably flawed."

Over the years, my dad wrote about gravitation for many different audiences, from interested laymen and potential investors to the most sophisticated physicists. You will encounter many of these essays on the pages ahead, but even the simplest of them doesn't meet the *paper-napkin* standard. I was the beneficiary of the paper-napkin lecture on gravity many times in coffee shops and restaurants from Seattle to Palo Alto between 1975 and 2012, but much to my regret, I never wrote any of them down. I was always "too busy." Now, however, I will try to reconstruct a paper-napkin lecture for *you*, because without it, much of what follows will seem more difficult than it really is.

A word of caution to experts in general relativity, or those who fancy themselves experts: You won't *like* what you're about to read. Your first instinct will be to dismiss it. Before doing that, however,

---

[21] *Waldo & Magic, Inc.*, 1950

please remember that F. Alzofon learned the special theory of relativity and the general theory of relativity at Cal Berkeley from physicists who were colleagues and friends of Albert Einstein. One of them was an authority on relativity lauded by Einstein himself. Consequently we may surmise that Dr. Alzofon knew at *least* what you know about relativity, and probably had far more grounding in its historical origins. Not surprisingly, he could handle himself more than adequately in a debate with experts, though such debates were few and far between, because time and again, the experts showed that they could not stomach a critique of their basic assumptions, most of which they had forgotten or taken for granted in the long road toward mastering the intricacies of the GTR. Such mastery was intimately connected with their livelihood, and, as observed by Upton Sinclair, "It is difficult to get a man to understand something, when his salary depends upon his *not* understanding it."

Their behavior was rather stereotyped: When they felt the ground begin to shift beneath their feet, they flew into a rage and ran for the shelter of the ivory tower. This happened not once, but over and over. Beyond "behavior unbecoming to a scientist," their rage resembled a scene from a *Far Side* cartoon by Gary Larson. Not pleasant to hear, I'm sure, but quite true, unfortunately.

So if you find what follows unpalatable—and I'm anticipating you will—please reserve judgment and then *read the papers* listed on page 259 and elsewhere in the book. They speak at your level of sophistication. And remember, the UFT has *always* been open to experimental refutation or confirmation. The outline of such an experiment is given here, beginning on page 131. So why not pick up the gauntlet? Nothing in it for you, after all—except a few weeks' work and a Nobel Prize.

THE PAPER-NAPKIN VERSION OF GRAVITATION

One thing I learned when trying to repeat my father's concepts back to him: A scientific argument is extremely precise, and the whole edifice might stand or fall on misuse of a single word. For that reason, I will begin each step in the argument that follows with a quotation from one of my father's papers or from our conversations. I will then give a paragraph or two in plain English that will explain the implications of what he said. I may go astray in the second part, but not in the first.

1) "One fundamental feature of the unity of nature is that only two forms of the reality observed exist: radiation and matter."[22]

Observable reality comes in only two flavors: matter and radiation. That's all he's saying. Take a look around you: Can you see anything that's not either radiation or matter?

"Radiation" means electromagnetic radiation, which ranges from visible light through the invisible spectrum of infrared and ultraviolet, on through gamma rays, X-rays, and more. It's a broad spectrum,

---

[22] F.E. Alzofon, "The Unity of Nature and the Search for a Unified Field Theory," p. 600; *Physics Essays*, *6* (1993) 599-608. (Reprints available through *Physics Essays*. Used by permission; see note on copyright page.)

but it's all the same thing, just electromagnetic waves of differing frequencies.

"Matter" means solid objects, ranging from planets, stars, and galaxies on the macroscopic scale down to everyday objects, such as tables, chairs, and the ground we walk on, and, diving deeper, the invisible, microscopic world of molecules, atoms, and elementary particles. At base, *everything* material, from galaxies to table salt, is composed of elementary particles, which includes electrons, neutrons, protons, mesons, bosons, quarks, and so on. As with radiation, it is a broad spectrum, but all the same thing, in essence. One cannot subdivide elementary particles further without finding that they are composed of condensed radiation. This is the meaning of Albert Einstein's famous equation $E = mc^2$:

"This relation [$E = mc^2$] is a way of saying that radiation and matter are equivalent: that is, radiation is dispersed matter, and matter is condensed radiation. On a subatomic scale, this assertion has been proved experimentally: one can be transformed into the other."[23]

Here's the point: While the real world is full of infinite variety, it is only one of two things at the root: matter or radiation. And at some level, these two are equivalent.

"It's all one thing, man."

My dad didn't say that. Actually I think it was a hippie stoned on acid I met in Golden Gate Park in 1967. But he wasn't wrong. All reality is a single unified field composed of electromagnetic energy.

Now this is all very fascinating, but what does it have to do with gravity? Simply this: It implies that we should be looking for the *physical* cause of gravitation in some property of the unified field of matter-radiation. This is a commonsense assumption,[24] but it is not the same assumption that guides general relativity, the currently accepted model of gravitation.

There *is* one thing that all matter has in common, and it was stated above: *elementary particles*. This leads to quote No. 2:

2) "So the question, of course, is what property do elementary particles have in common that will give rise to the gravitational force for all of them? Because the thing is, that the gravitational force appears to be the same for any type of mass—for stars interacting with the Earth, or with each other, or cold dust in space, over a wide range of temperatures and types of matter: You have the same force acting, and the same force law. With respect to our modern knowledge, we know there are many kinds of elementary particles and yet there again, there seems to be the same force acting, the gravitational force, no matter what kind of elementary particle there is."[25]

---

[23] See *Chapter 34*, p. 214

[24] It is also *logical positivism*. See pp. 183, 231, 246.

[25] See *Chapter 31*, p. 194

A-ha! Gravitation depends on only one thing: *mass.* The composition, temperature, or physical state of the mass has no bearing. Only the *quantity* of mass. And what does all mass have in common, regardless of its composition, temperature, or state? *Elementary particles.* It stands to reason then that we should look to some property of *elementary particles* for the source of gravitation.

3) "Macrocosm reflects microcosm," and conversely, "Microcosm conditions macrocosm."

This quote comes from a telephone conversation I had with my father sometime in the early 2000s. It implies that the behavior of big things, such as galaxies and planets, reflects the behavior of small things, namely elementary particles, because that's what the big things are made of—elementary particles, and nothing else. The second phrase implies that the microscopic world of elementary particles influences cosmic phenomena, such as planets, moons, stars, and galaxies. Put another way, "The roar of a freeway is the sound of a single automobile multiplied a thousand fold." Or, conversely, "A drop of water determines the character of an ocean."

Surely you can think of a few more Zen-like sayings such as the above. Why not give it a try?[26]

Now take another look at quote No. 2 above:

"What property do elementary particles have in common that will give rise to the gravitational force for all of them?"

In other words, if we can discover an attractive force between *two* elementary particles, then the same force will exist for *all* elementary particles. On a microscopic scale, the force may be very small, but in aggregate, it will be very large. *Microcosm conditions macrocosm*: It *becomes* the gravitational force.

4) "We imagine the [elementary] particle to be made of a dense accumulation of energy, which is increasingly dense toward the center, less so as you go out toward the outside, until the matter shades off into a field—the gravitational field. I view the field—the gravitational field—and the matter composing the particle as a single entity."

Understanding the structure of a particle is the first step in tracing the cause of the attractive force we call gravitation. The words above can be visualized in the diagram below:

---

[26] The "microcosm-macrocosm" saying sounded too much like the pre-Socratic Greek philosopher Heraclitus or the Chinese philosopher Zhuang-Zhou (Chuang Tzu) for me to believe that my dad made it up, so I did a Google search, and came up empty: Apparently he *was* the first. But this was one of his most outstanding talents, a knack for generalization, or in scientific terms, *inductive reasoning*. He considered science to consist of an alternation between induction and deduction. We see this protocol at work in the paper-napkin argument and in his papers.

This is your first "paper napkin sketch." It depicts the *energy* in an elementary particle, and *not* the actual *shape* of the particle. Many possible shapes have been proposed for particles, but that doesn't concern us here. We *only* care about the energy *density* (the height of the curve) as it relates to particle *diameter* (measured on the baseline scale). As you can see, the energy is "increasingly dense toward the center," less so toward the edges, and shades off toward infinity on the fringes.

My dad studied particle physics at UC Berkeley as part of his PhD program and performed experiments at the cyclotron, the Lawrence Radiation Laboratory, or what my mom and dad used to call "The Rad Lab." He published in *Physical Review* on the topic, so the view he is taking of the energy density of a particle, while simplified, is sophisticated and well-supported by experiment.

What is *new* is his view of the gravitational field as part of the extended field surrounding a particle. This implies a difference in composition of the field near the center of the particle and on its fringes. Apparently he viewed the activity on the fringes of a particle as a loosely bound halo of random electrodynamic activity. You might think of it as a "blizzard" of electromagnetic vortices surrounding each and every elementary particle. Within the blizzard, charged particles and photons come into existence and disintegrate with extreme rapidity.[27] The name for this random noise is *virtual processes*.

Any talk of virtual processes takes us out of the realm of "napkin sketches" into jargon country. For this reason, I will footnote a few quotes from his papers and leave it at that.[28] As far as the general reader is concerned, the thing to focus on is the bell-curve variation in energy density shown above. With this easily visualized model in hand, we are on the brink of solving the riddle of gravitation.

5) "The extended field surrounding the particle is asserted to be responsible for the gravitational force. To see how this occurs, consider what happens when two elementary particles are neighbors, so that their extended fields overlap."[29]

---

[27] 4.3×10^{-10} seconds, or about two billion times faster than the fastest gun in the West! (see 1981 paper, p. 4)

[28] See p. 106

[29] *Chapter 34*, p. 215

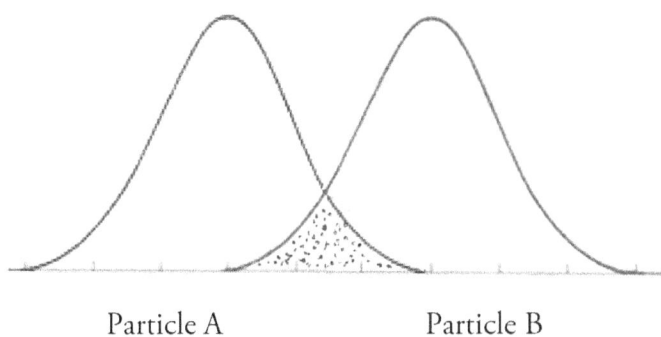

Particle A   Particle B

Here is napkin sketch No. 2, and what does it show? The fringes of two particles overlapping in the shaded area. Elsewhere, my dad points out that when mass is added to a particle, the binding forces holding it together must increase. As far as Particle A is concerned, its mass has increased by the amount shown in the shaded area. The same is true for Particle B on the right. The mass increase causes the particles to draw closer together. Here's how he puts it, minus the math:

6) **"Insofar as one particle is concerned—say, the one on the left—it has gained in energy by a small amount (...), and the condition for stability has altered (...), [resulting in] a tendency to draw the added energy closer to the central portion of the particle. The second particle (on the right) senses this tendency as a force upon it as a whole. When multiplied by the enormous number of particles making up macroscopic matter, this minuscule effect becomes, in aggregate, what we call *the gravitational force.*"**

And there it is: Gravity! *Minuscule* and weak on the microscopic scale, *large* and irresistible on the cosmic scale. But even after hearing this explanation a hundred times, I still asked my dad repeatedly, "How does gravity act at a distance?" In other words, how does the Earth pull on the moon and vice versa? What is the physical connection between the sun and the planets? The answer was staring me in the face the whole time. Here it is:

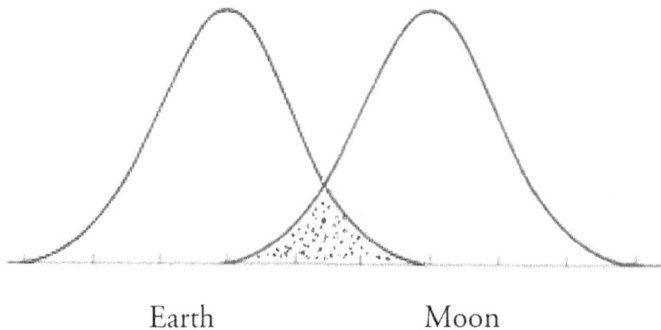

Earth   Moon

By repeating the napkin sketch of Particle A and Particle B, I am emphasizing that macrocosm reflects microcosm. The sound of a freeway is the sound of a single car multiplied a thousand fold. Or similarly, the attractive force of a gazillion elementary particles overlapping their extended energy fields adds up to the force of gravity.

As you might imagine, justifying such a simple idea to a crowd of skeptical physics professors is no trivial task. That's why the bulk of my dad's formal papers were consumed with topics such as a reinterpretation of Heisenberg's uncertainty principle,[30] "The Dimensionality of Space-time and Spontaneous Processes,"[31] "The Equations of Motion and Quantum Mechanics,"[32] "The Relation of the Matter-Radiation Field to Classical Fields and Elementary Particle Force Fields,"[33] and "The Zero-Point Energy of the Vacuum"[34]—lots of sophisticated math and lots of subjects only a physicist could love. You can get a taste of this in *Chapter 37*. If you're a physicist, you might enjoy it.

But this is the napkin-sketch version of gravitation, not a PhD dissertation, so let's leave the tire-kicking to the experts and take a brief look at gravity control in light of our theory.

THE SECRET OF GRAVITY CONTROL

Notice that we have found the crossover point between electromagnetic forces and gravitation. It is the random electrodynamic activity of virtual processes that surrounds all particles, and by extension, planets, moons, stars, and galaxies. But how can we use our control of electromagnetism—which by now is rather sophisticated—to influence the random, extremely short-lived virtual processes associated with gravitation?

My dad was a world-class expert in heat conduction (see *Appendix D*, p. 263), so naturally he leapt upon the analogy between heat, which is the random motion of molecules, and gravitation, which is a byproduct of random fluctuations in the matter-radiation field (i.e., virtual processes). He pursued the analogy in depth in *Appendix A* of the 1981 paper, under the subtitle "Kinetic Theory Model of the Gravitation Field," and in "The Unity of Nature," p. 603. References to the analogy can be found throughout the present volume on pages 105, 106, 187, 216, and fn 1 (p. 17).

But we don't need to add the heat analogy to our paper-napkin lecture. All we need to note is the solution: *If gravity is analogous to heat, gravity control is analogous to cryogenics.* Cryogenic cooling reduces the energy in randomly jittering molecules; gravity control reduces the energy in randomly fluctuating virtual processes.

---

[30] "The Unity of Nature," p. 605

[31] 1981 AIAA paper, p. 10

[32] "The Unity of Nature," p. 604

[33] 1981 AIAA paper, p. 17

[34] "The Unity of Nature," p. 606

The way this works is a two-step process: **Step 1)** We *order* the virtual process cloud of the vehicle through a clever recipe of electromagnetic fields. **Step 2)** We *shut off* the electromagnetic fields and let the Earth's virtual process cloud, which is entirely random, expend energy battering the ordered cloud of the vehicle back into randomness. The order-disorder, on-off cycle is applied over and over until the Earth's gravitational field weakens to the point of almost zero in the vicinity of the vehicle. The technique is much like a well-known cryogenic method that easily achieves temperatures close to absolute zero (zero degrees Kelvin). See *Appendix B*, p. 255, for more.

The reason gravity control consumes relatively small amounts of energy is because, like the cryogenic process, it works like a pump: each time the on-off cycle is repeated, a little order is stored, while the "temperature" of the gravitational field is lowered a little more. The weight loss adds up until all the weight gone. My dad predicted that at a certain point a "cascade effect" would occur, and the weight of the object would rapidly fall to "absolute zero." This effect *was* observed in the 1994 experiments.

CONCLUSION

This has been the "quick-and-dirty" introduction to gravity and gravity control. What it lacks in supporting evidence and sophistication, it makes up for in clarity, clean lines, and common sense. You don't have to look under the hood of a Porsche to appreciate its beauty. Similarly, you can appreciate the elegance of the theory and applied technology we're offering without expert knowledge. This broad introduction will serve you in good stead as we swim into deeper waters ahead.

FOOTNOTE 28 (page 103): From *Appendix B – A Notebook Fragment*, p. 255: "The magnetic moments of the nuclei and electrons are conceived as consisting of two parts: a) A hard core of dense, circulating electric currents, essentially responsible for the usually-measured values of the magnetic moments. **b) A cloud of electric charges due to the virtual processes, which are not as tightly bound to the hard core as the currents making *up* the core.**"

From "Unity of Nature," p. 600 (used by permission; see note on copyright page): "One of the experimental properties of radiation in a black-body cavity in thermal equilibrium with the walls of the cavity is the increasing magnitude of fluctuations when radiation is confined to smaller and smaller regions of space-time. Such a property need not be true of a light signal propagated in a vacuum under nonequilibrium conditions, but the existence of such a possibility is central to quantum electrodynamics. The accuracy of its predictions argues that virtual processes (these are equivalent to fluctuations) do indeed exist. The physical model associated with virtual processes was essential in the development of the theory proposed below,[3-5] but the writer has found that experimental physicists often do not believe in the reality of virtual processes, since they arise in connection with a perturbation theory that does not have universal validity. However, the physical model used in the development of the theory proposed can be established on a classical basis which has thus far been shown to be equivalent to quantum electrodynamics."

Chapter 18

The 1989 Grant Proposal

EDITOR'S INTRODUCTION

The following is a composite of material from three drafts of a proposal that was ultimately sent to a nonprofit research foundation. The '89 proposal was written after the DoE proposal, which has been lost, and no doubt contains improvements. The idea behind combining material from all three drafts was to present the best from each.

The proposal emphasizes new forms of energy production because that was the foundation's primary area of interest. The discussion follows a problem/solution format: First, the physics problem is broadly stated, then the focus shifts to gravitation and how gravity control research will improve the outlook for clean energy.

The general problems in our current understanding of gravity are outlined, and the UFT is offered as a credible solution. The rationale for the experiment follows, with a much more detailed description of the technology than anything presented so far in *Book II*. A list of required equipment follows, with prices updated for the present day (2017).

Chapter 19

Grant Proposal (August 14, 1989)

A Program to Develop New Sources of Useful Energy:
Phase I – Controlling the Gravitational Force

INTRODUCTION

The following proposal will concern Phase I of a program to develop new methods of generating useful forms of energy. The primary distinguishing feature of this program is the use of a unified field theory providing a new model of the physical universe, which is formulated in terms of easily visualized and accepted concepts that are readily amenable to experimental tests.

One such test is the use of the theory and its physical model to originate devices for the control of the gravitational force. Phase I of the program will concern an experiment to demonstrate the feasibility of altering (weakening) this force. The discussion and analysis that follows presents a rationale, schedule, and costs for the experiment. Some practical uses of the process are indicated.

GENERAL TOPICS

• **Energy Generation**

A problem of great general and topical interest is the discovery of new forms of useful energy generation. The publicity given to recent claims of accomplishing cold fusion,[35] and the ready appropriation of large sums of money by a state legislature for future research is a demonstration of public perception of the importance of this question.

---

[35] In March, 1989, University of Utah electrochemist Martin Fleischmann reported production of heat by nuclear processes at room temperature. The press called it "cold fusion." By late 1989, cold fusion was discredited in the minds of most scientists, who lambasted Fleischmann's methods. At the time my father wrote, the bad publicity was yet to occur. The uproar, however, indisputably demonstrated public interest in energy production.

• **Gravity Control**

The second problem to be discussed is the control of gravitation. In the discussion to follow, control of gravitation will refer to alteration of the physical mechanism producing the gravitational force. It does not mean the introduction of a force in addition to the gravitational force that may be directed either opposite to, or in the same sense as the gravitational force.

BACKGROUND

• **Energy Production**[36]

Since energy cannot be created or destroyed, the "production of energy" is a term which must refer to the conversion of one kind of energy into another, and useful, form of energy. Often the conversion takes the form of transforming a potential energy into a useful kinetic energy. For example, potential gravitational energy is transformed into electrical energy by use of falling water, or a gas under pressure may expand to cause motion of a piston.

Approximately characterized, the history of energy generation has progressed from an original dependence on natural sources such as muscle and wind power, etc., to use of exothermic chemical reactions, then to production of electrical power, and finally to production of subatomic reactions (nuclear reactors, fusion, etc.). In general physical terms, this progression has been toward reliance on physical processes involving energy stored in smaller and smaller entities and greater power yields. But in each case, finding a useful energy transformation depends on knowledge of particular details of a specific process which makes the desired energy conversion possible. Only experiment can disclose these details; however, experiments cannot proceed without an adequate theory to guide the way. This theory must tell how to convert the random (or probabilistic) motion of atomic and subatomic entities to macroscopic, ordered motion for extraction of useful work.

At this time, further progress in employment of subatomic processes for extraction of power is hampered by lack of understanding of subatomic processes. Although much is known from experiment, the theory of such processes is admitted to be inadequate. This lack of complete understanding is illustrated by the numerous infinities that result from modern theory, e.g. the concept of an infinite amount of energy stored in a "vacuum."

It has been hoped that discovery of a unified field theory will resolve the difficulties inherent in present models of reality; it is the writer's contention that his unified field theory accomplishes this role, and will show how the desired energy conversions can occur.

---

[36] This section referenced on p. 32.

• Gravity Control

The usefulness of gravity control is easier to understand at present than before the space exploration programs began. It is certain now that anyone who can offer a practicable scheme for controlling gravitation is certain to profit from government contrasts, at minimum. In addition, civilian organizations are certain to follow suit, leading to extensive space colonization shortly afterward. The latter will be desirable to relieve population pressure on resources, for example. It is also expected that practicable reduction of the weight of large objects of commercial value will make it possible to transport them without fitting them into containers to be moved on common carriers.

At present the launch of rocket vehicles is strongly dependent on wind and weather because rockets are not readily maneuverable at the high launch speeds (i.e. escape velocity) required with vehicles propelled only by rockets. With gravity control, the amount of rocket fuel carried by these vehicles can be reduced and the launch velocity can also be reduced, leading to greater maneuverability of the vehicle; this same maneuverability will also lead to more easily accomplished exploration of the solar system and planetary bodies. With use of a nuclear reactor, materials found in space can be used for vehicle propulsion, greatly extending the operating range of the vehicle. In addition, it will then be possible to accelerate during the first part of any trip, and decelerate during the second half of the trip, materially reducing the total time required for the trip.

GRAVITATION

• Choice of Model

At present, no clear model of the physical processes giving rise to the gravitational force is generally accepted. Since the unknown provides an open invitation to speculation, a great number of theories have been advanced to explain the origin of the gravitational force. Many of these theories claim the possibility of eventual control of this force. How, then, can anyone distinguish between promising and unpromising proposals when so many kinds are offered? I believe the criteria of the scientific method provide a good and practical guide to how to choose a useful line of investigation.

• The Scientific Method

There are several definitions of what science is. For example, mathematics is often claimed to have no relation to the real world, but it is agreed to be a science. I am here, however, limiting the discussion to physical science, for which experiment, as well as theory, are necessary complements. With this understanding, the scientific method, as Galileo and Newton would have defined it, is comprised of the following steps:

1. A correlation between physical events is observed in Nature.

2. A theory is proposed to explain how these correlations occur.

3. The theory is tested by experiment.

These are hardly enough to describe how physical research is practiced, although the above three steps are often the only three listed. In addition, we must add the following criteria:

4. The theory must provide *numerical* predictions; experiment must verify them *numerically*.

5. The theory must contain the successful theories of the past as special cases.

6. The concepts used by the theory must lend themselves in a convenient manner to experimental design and interpretation. The scientific method is a *social* enterprise.

7. The theory must be easier to use than competing theories.

8. The theory must be falsifiable, that is, *refutable*: It must be possible to conceive of an observation that would prove the theory, or a hypothesis derived from the theory, false.

The writer has formulated his unified field theory to conform to the above requirements for a scientific theory; experiments are now needed to test it.

• Identification and Significance of the Problem

The force of gravitation is ubiquitous: on Earth and everywhere in the Universe. Nothing can shield matter from its effects. The strength of the force is steady, strictly proportional to mass and independent of the chemical composition or state of matter. It influences the growth of plants and animals in an essential manner and is required for the continued health of the human body. It has to be taken into account in every task performed in everyday life, as well as in the manufacture of high-technology products. Heavy industry makes use of it without much thought given to its necessity (e.g. in pouring molten metal). Space exploration and placing satellites in orbit are largely a matter of overcoming gravitational attraction for the vehicle being used; Newton's Universal Law of Gravitation is routinely used to plot orbits for exploration of the planets. The same Law is also useful to the military for predicting the course of artillery shells. Other uses can be conceived.

Yet, in all of the above applications of the understanding of the role of the gravitational force in human affairs, there is no perception of how the gravitational force arises, of what physical processes originate the force of gravitation. As a consequence, severe limitations are placed on all enterprises that must take this force into account. For example, very large quantities of rocket propellant must be used to free any vehicle from the Earth's gravitational field, and, as a consequence, there is little propellant remaining for maneuvering within the field in case of bad weather or for unforeseen circumstances in a trip to the Earth's moon and return. Similar difficulties prevent extended exploration of the planetary system by manned rocket vehicles; the expense and awkwardness of mission planning rises rapidly with the distance to be traveled. Colonization of the Moon and the planets is severely hampered by such considerations. Understanding the origin of the gravitational force and its control can be expected to facilitate space exploration and exploitation. In addition, closer to home, we can expect that completion of tasks on the Earth's surface which may be hindered by the gravitational force can be facilitated by an understanding of its origin and control.

• Attempts to Discover the Source of Gravitation

The gravitational force is so omnipresent that it was taken for granted until Newton formulated his Law in the late 1600s. Previous to that time, the planets and stars were thought to be placed in the heavens and moved according to divine mandate. After the publication of Newton's Law in 1687 and its subsequent verification by many observations, many theories have been suggested for the origin of the force. However, none of these theories have been translatable into means of controlling the force, as has been the case for electric and magnetic forces.

The perception of the importance of this control has become more acute with increasing activities of world governments in space. For example, in the late 1950s the Soviet scientist Kirill Petrovich Stanyukovich claimed to have solved the problem of the physical mechanism for the origin of the gravitational force and claimed that, as a result, he could easily construct spaceships to navigate space. The U.S. Air Force, concerned that this might be true, funded a study of the state of understanding of gravitation on an international scale (i.e. not restricted to Soviet studies of gravitation) which was published in 1960.[37] It was the conclusion of the writer of the study, Dr. Maurice Garbell, that none of the theories described to that date, with the exception of one,[38] appeared to be capable of engineering realization, again reflecting a lack of understanding of the physical mechanism giving rise to the gravitational force.

The survey underscores the fact that despite great advances in the understanding of subatomic phenomena, and of cosmology, gravity has stubbornly resisted a cause-and-effect explanation in terms of known physical phenomena.

Since the publication of Dr. Garbell's survey in 1960, increased emphasis has been placed on the role of a unification of gravity and other forces, and, indeed, if the objective is to eventually control the force of gravitation, it will have to be done through its dependence on other phenomena and forces. Moreover, in the contest of modern physical theory, in order to be convincing, this dependence must be stated in terms of a theory that is relativistically covariant, i.e. containing physical laws that have the same expressions for every acceptable coordinate system: this is required by the special theory of

---

[37] Ref. 1, p. 126

[38] Ref. 2 and 3, p. 126

relativity. The latter property is necessary in order that relative motion not introduce effects that are essentially new; for example, an electric charge that generates a purely electrostatic field when at rest relative to an observer, will be observed to also generate a magnetic field when in relative motion with respect to an observer: this feature must be included in any credible theory.

To summarize: a frequent theme in the attempt to explain gravitation in terms of familiar concepts is the effort to construct a relativistically covariant theory which describes a unified field: a field which relates the gravitational field to other, well-known fields. Thus, such a theory should include what is known about the electromagnetic field, gravitational field, quantum mechanics, the equivalence of matter and radiation, and can be extended to subatomic phenomena. When such a theory exists and also indicates how gravity can be controlled, it is necessary to perform experiments to verify its predictions.

It is asserted that such a theory now exists.[39] Moreover, since it includes classical and relativistic particle mechanics, electromagnetic fields, the classical gravitational field, a clearly visualized model of inertial and gravitational mass, the equivalence of matter and radiation, quantum mechanics, and a means of extending the theory without infinities to subatomic processes, the theory has been verified in part by thousands of experiments, and therefore *must be considered to be an extension of present technology, and already partially verified by experiment.*

Moreover, the basic ideas are not new but well accepted, although they are put together in an essentially new way. In addition to the foregoing, we must consider the theory under discussion to be *more realistic than presently accepted theories* since the potentials of the fields characteristic of the theory do not possess any infinities, e.g. like the Coulomb potential and the static gravitational field (Newton's Law). Another, more realistic feature of the theory is the lack of an infinite zero-point energy.[40]

In sum, since the proposed theory agrees with most of present technology, and is more realistic than much of modern theories, there is an excellent chance that those parts of it that have not yet been tested by experiment will *also* be relevant to reality and correct. This is the opportunity offered by the proposed program. A further indication of the importance of this opportunity has been afforded by the remark of the writer of *Soviet Research on Gravitation*,[41] that, "Out of all the theories reviewed, [i.e.

---

[39] Refs. 2, 3, 4, 5, p. 126

[40] A valuable supplement to the above can be found in Ref. 3, p. 126, from p. 1, col. 2, para. 1, "Although the above deductions...," through p. 2, col. 2, para. 2, which begins "We here depart from Einstein's geometrical formulation...." The lengthy excerpt was not included out of respect for the AIAA copyright.

[41] Ref. 1, p. 126

to the date of the publication] the theory advanced in *The Origin of the Gravitational Field*[42] is the only one that appears capable of engineering realization."

PHYSICAL BACKGROUND – OVERVIEW

• Origin of the Gravitational Field

Briefly summarized, the unified field theory model[43] holds that the origin of the gravitational field is due to fluctuations in forces which are already known to exist (e.g. electromagnetic forces) and to the stability of elementary particles comprising ordinary matter. The UFT is discussed in more detail in the next section below.

• Method of Altering the Gravitational Force

By analogy to the case for the adiabatic cooling of paramagnetic salts, it is expected that the force called into play by the latter fluctuations should be reduced by compelling the fluctuating field to do work against oriented nuclei and electrons of suitably chosen materials. The orientation can be accomplished by dynamic nuclear orientation utilizing a constant magnetic field and a microwave field applied to the specimen. Alternately applying and cutting off these fields is expected to reduce (or, with different timing, increase) the gravitational force on the specimen. A more complete rationale for the experiment is given below.

THE UNIFIED FIELD THEORY MODEL

In order to understand the rationale for the experimental design to be proposed, it is necessary to discuss the model to be employed for visualizing the origin of the gravitational force.

The theory in Refs. 2, 3, and 4 emphasizes the role of fluctuations in the energy density in the region (assumed to be a vacuum) between any two objects exerting a gravitational force on one another. The gravitational field is conceived to be generated by these fluctuations, although the field, on a terrestrial scale of measurement, appears to be unchanging in time, i.e. it is a static field. Thus, the average effect of the fluctuations is Newton's Law of Gravitation:

$$F = \sqrt{G}\ M\ \sqrt{G}\ m/R^2 \text{ (dynes)} \tag{1}$$

where G is the gravitational constant (6.67 x 10^{-8} dyne cm^2 gm^{-2}), M and m are the masses of the interacting bodies, and R is the distance between them (cm). For our purpose, the distance between

---

[42] Ref. 1, p. 126. Ed note: Private communication from M.A. Garbell to the writer (FA). Ref. 1 corroborates what FA wrote here, but less directly, by alluding to the possibility of altering the gravitational force using Alzofon's theory; Garbell says nothing of the kind about any other theory, including, of course, the GTR.

[43] Refs. 2, 3, 4, p. 126

the objects is considered to be measured between the centers of spheres, if the objects are extended bodies (e.g. such as the Earth), or to the point at which an elementary particle is located (assumed to be, essentially, a point mass). The force F is exerted by one object upon the other and is attractive.

Briefly summarized, the origin of the gravitational field is held to be due to fluctuations in forces that are already known to exist (e.g. electromagnetic forces) and to the stability of elementary particles comprising ordinary matter.

The basic premise of the model is that, since the electromagnetic field is observed to undergo fluctuations in intensity, and since it is the determinant of the space-time metric, and cannot be described in terms of itself, this feature of the field must be included as part of the structure of space-time. When the mean motion of a feature of the field is zero, nothing but fluctuating motion remaining, this constitutes the definition of a mass particle. Moreover, since the particle is described by a field, it is not localized at a point (like the Newtonian particle, or the point masses described above), but owing to its stable existence, most of the field is localized to a region of the diameter of a Compton wavelength: h/mc, where h is Planck's constant (6.6 x 10^{-27} erg sec), m is the mass of the particle in gm, and c is the speed of light (3.0 x 10^{10} cm sec^{-1}). If two such particle fields interact, the fields overlap and any given one of the particles gains energy and therefore mass: the mass increases, say, from m to $m + \Delta m$. As a consequence, the region in which most of the mass-energy is concentrated becomes $h/(m + \Delta m)$, or approximately $(h./mc)(1 - \Delta m/m)$, i.e. less than before. Since the particle has no way of distinguishing between its original mass energy and that contributed by the other particle, it tends to contract and to pull on the energy contributed by the other particle, drawing the other particle toward it in the process. This is an example of a general principle, *Le Châtelier's Principle*, which asserts that if any physical system is in equilibrium and there is a small change in its state, then the system will alter in such a way as to restore the state of equilibrium. In this case, the given particle acts to restore a condition in which the new Compton wavelength characterizes the diameter of the central region of mass-energy.

A further consequence of the unified field theory is that no fluctuations of the energy in a vacuum need be invoked; there is no infinite quantity associated with the zero-point energy in a vacuum. Indeed, there is no infinity associated with any quantity appearing in the theory. For example, the force F in equation (1) tends to infinity as the distance between the interacting masses tends to zero; in the unified field theory, Newton's Law can be derived in the form (1), but it is shown that as R tends to zero, the force law changes to a different form that does not become infinite as R tends to zero. Moreover, the property of inertial mass is itself shown to be a consequence of the fluctuating motion of the field: to the extent that the field participates in the fluctuating motion, to precisely that extent the field has the property of mass inertia. In addition, as a result of this model, the notion of "virtual" charges $\sqrt{G}\,M$ becomes understandable as real charges and an integral part of the unified field. For example, since the energy density of the gravitational field is given by:

$$(1/8\pi)\,(\sqrt{G}\,M/R^2)^2 \text{ ergs cm}^{-3} \tag{2}$$

(derived in a manner analogous to the electrostatic energy field density), we may estimate the number of equivalent charged particles in the field at the distance R from the source mass M, assuming most of them to be electrons (since these require the minimum energy in the creation-annihilation process for *particles*; photons require less minimum energy):

$$(1/mc^2)\,(1/8\pi)\,(\sqrt{G}\,M/R^2)^2 \text{ ergs cm}^{-3} \text{ particles} \tag{3}$$

where m is the mass of the electron (9.1×10^{-28} gm). Thus, at the Earth's surface (R = radius of the Earth = 6.4×10^8 cm), the energy density is about 5.7×10^{11} ergs cm^{-3} to one significant figure. Since the rest energy of an electron is 8.2×10^{-7} ergs, an order of magnitude estimate of the number of charges in the Earth's gravitational field at the Earth's surface is about 7.0×10^{17} electron charges per cubic centimeter, also to one significant figure. *These are equivalent charges and do not have a long-term existence; they are a consequence of creation-annihilation processes.*

RATIONALE FOR THE EXPERIMENT

Since the existence of the gravitational field is identified with the existence of the fluctuating motion of the energy in the field, then reduction in the fluctuating motion should reduce the intensity of the field and reduction in the force F in equation (1).

An analogous situation exists in the use of a magnetic field to reduce the temperature of a paramagnetic salt. In the latter process, a magnetic field (constant) is applied to a specimen of the salt, orienting the elementary molecular moments of the salt. After thermal equilibrium is established, the magnetic field is suddenly removed, and the immediate surroundings do work on the oriented magnetic dipoles, losing some of their random motion in the process, while causing the dipoles to lose some of their orientation, and thus causing a temperature drop in the neighborhood of the salt.

We propose to reduce the random motion in the gravitational field in an analogous manner by use of dynamic nuclear orientation, a well-established technique of physical research.[44]

In principle the method of dynamic nuclear orientation is easy to state; its purpose is to orient nuclei

---

[44] Ref. 6, p. 126

of atoms instead of molecules of a paramagnetic salt, but the method of orientation is similar. Thus, a constant magnetic field is imposed on a specimen of a ferromagnetic material, causing the electrons of the atoms to precess about the direction of the field with a characteristic (Larmor) frequency. A relatively weak magnetic field which varies at the Larmor frequency is then applied to the specimen, causing the electrons to tip over and become oriented. To preserve the angular momentum of the specimen, the nuclei must also tip over and become oriented. This process usually is carried out at liquid helium temperatures to eliminate the effect of thermal disruption of the orientation, so that studies of nuclear properties can proceed without interference from many thermal collisions with surrounding atoms. For ferromagnetic atoms, once the oscillating magnetic field is removed, the orientation decays more or less rapidly, depending on the temperature of the heat bath in which the atoms are immersed; for room temperature, the decay time is less than a microsecond, so that the experimenter cannot perform any of the experiments we have in mind with such conditions.

Fortunately, a device can be used which results in longer-term nuclear orientation at room temperature. It has been found that if ferromagnetic atoms are thinly embedded in a rare earth element (e.g., iron or chromium embedded in aluminum or magnesium) the ferromagnetic nuclei can again be oriented and will impart their orientation to the rare earth nuclei. In turn, it is found that the orientation of the rare earth nuclei will last for as much as 6 milliseconds after the oscillating magnetic field has been removed.[45]

The above considerations indicate how an experiment designed to show how the force of gravity can be altered (i.e. weakened) can be performed:

A small specimen of aluminum (very pure to avoid interfering resonances from isotopes) with a small amount of colloidal iron powder inserted into it (easy to do since the melting points of aluminum and iron are so different) is placed in a magnetic field of about 2000 oersteds. A microwave field is applied to the specimen at the Larmor frequency (determined by the strength of the magnetic field) and observations are made to see whether or not the weight of the specimen has been altered. There must be a means of verifying that resonance of microwave and natural electron precession frequency is maintained: this will be a critical part of the experiment. It is proposed to sample the resonance at 6 millisecond intervals; indeed, the microwave field will be turned on and off at 6 millisecond intervals, each microwave pulse to last for a few microseconds to preserve nuclear orientation. The latter proposals are based on the following observations. In these observations, we shall refer to the fluctuations in energy (including creation and annihilation of radiation photons and of particle) as "virtual" processes. The term was common in the quantum electrodynamics where these processes were thought by some physicists to be a convenient formal device, although at least one paper advanced

---

[45] Ref. 6, p. 126

the opinion that they were real.[46] *Ref. 8*, by Dr. H. E. Puthoff, also emphasizes the reality of such processes.

From observations on gravitational fields, it is evident that "virtual" processes have the following properties:

a) Since the net effect of two gravitational fields at a given point is simply the sum of the two fields (i.e. they do not interfere with one another), it follows that "virtual" photons/particles interact very weakly.

b) "Virtual" photons/particles interact strongly with matter.

c) Any given mass does not recognize a distinction between the "virtual" photons/particles it generates and the photons/particles that a mass interacting with it generates.[47]

THE PHYSICAL MECHANISM FOR THE REDUCTION OF GRAVITATIONAL FORCE

The reduction of the gravitational force by the means proposed above is imagined to occur in the following manner:

Upon orientation of the aluminum nuclei and electrons, the "virtual" processes owing to both the Earth and the aluminum nuclei and electrons give rise to charged particles which are oriented owing to their strong interaction with the oriented magnetic moments. This gives rise to a contribution to the net magnetic moments of nuclei and electrons; in the case of the electrons the additional magnetic moment has been calculated by the methods of the quantum electrodynamics and found to generate the fractional alteration:

$$\frac{\mu_e - \mu_0}{\mu_0} = 0.0012 \qquad (4)$$

where the subscript o refers to the original magnetic moment of the electron, and the subscript *e* refers to the magnetic moment including the effect of creation and annihilation of charges. We shall assume that the same relation applies to the magnetic moment of the aluminum nuclei, for lack of a better value; since the magnetic moments of aluminum nucleus and electron are in the ratio of 5 to 1,[48] it is likely that a greater fractional alteration occurs for the nucleus.

The frequency of the microwave field used for tipping the nuclei is given by:

---

[46] Ref. 7, p. 126

[47] Ref. 3, 4, p. 126

[48] Ref. 9, p. 126

$$\nu = g\mu_B \, H/h \tag{5}$$

where g is the spectroscopic splitting factor for the electron, μ_B is the magnetic moment of the Bohr magneton, and h is Planck's constant. Typical orders of magnitude are $g \approx 3$, $\mu_B \approx 10^{-20}$ erg/oe. The alteration in the magnetic moment induces a change in the above frequency, with an attendant alteration in the energy given by multiplying the above relation by h. Using the value of the magnetic moment for a proton (i.e. $eh/2M_P$, where M_P is the mass of the proton), or 5.1 x 10^{-24} erg/oe, we estimate the effect of the alteration to be:

$$\Delta E = g \, (5.1 \times 10^{-24}) \, H \, (0.0012) \tag{6}$$

Since, for a paramagnetic ion, the nucleus experiences a local magnetic field of about 5 x 10^6 oe[14], $\Delta E \approx 9.2 \times 10^{-20}$ erg. The frequency interval associated with the latter quantity is then $\Delta E/h$, or about 1.4 x 10^7 Hz and the complementarity of frequency and time intervals implies a lifetime for the oriented "virtual" state of about 7.1 x 10^{-8} sec. It follows that the lifetime of the "virtual" states in question are very much shorter than nuclear orientation thermal lifetimes. A pulsed oscillating field will therefore serve to keep the paramagnetic nuclei oriented insofar as thermal decay is concerned and, properly timed, will allow that part of the magnetic moment due to "virtual" processes to become disordered.

The disordering referred to above occurs as follows: since, owing to the properties of "virtual" processes listed in (a) to (c) in the "Rationale for the Experiment" above (p. 117), a given nucleus cannot distinguish between those "virtual" photons/particles it generates and those contributed by, say, the Earth's gravitational field. It will therefore orient that part of the gravitational field's contribution in which it is embedded. As the cycle of alternate application and cutoff of the microwave field is repeated, a cloud of oriented "virtual" photons/particles is built up around the aluminum specimen and diffuses outward, limited by mean free path and lifetimes of the "virtual" states.

The constant magnetic field and frequency of the oscillating field can be selected to conform to the values often used in dynamic nuclear orientation. For example, with a fixed magnetic field H = 660 oe, one can employ a pulsed oscillating field of 3000 MHz, with pulses lasting 2 microseconds and a repetition rate of about 6 milliseconds. The cutoff of the oscillating field allows the cloud of oriented "Virtual" photons/particles to diffuse outward from the nuclei of the aluminum specimen, to be replaced by more oriented photons/particles, thus building up the extent, and the gravity-reducing capability of the cloud.

In accord with the model indicated in the foregoing discussion, the amount of e.g. iron, particles inserted into the aluminum metal is not critical, since its effects diffuse throughout the aluminum.[49] There are 2.2×10^{22} atomic nuclei per gram of aluminum, and each of these is the source of a reduction of about 9.2×10^{-20} ergs in gravitational energy. This is equivalent to a reduction of 2000 ergs per gram per cycle.

Since these cycles occur at the rate of about 10^4 per minute, we can expect about two joules per minute to be removed and stored in a cloud around the aluminum specimen. In these estimates, the rate of decay of orientation of the latter cloud, as well as the rate of orientation by the microwave field have been neglected. For the present these rates are unknown; the guiding rule for the experiment to be performed will be to select the experimental parameters so that at least ten percent of the weight of the specimen used is removed. For example, the rate of orientation is proportional to the square of the strength of the microwave field (i.e. amplitude of variation), and this is one of the parameters it is proposed to vary.

DESCRIPTION OF THE EXPERIMENTAL APPARATUS

This proposal is streamlined to deliver the maximum number of useful results with the minimum of investment.

In accord with this goal, the experiments will be performed at room temperature, although most of the dynamic nuclear orientation experiments have been performed at very low temperatures (tenths of a degree Kelvin above absolute zero). In addition, the generation of the microwave field is accomplished with the aid of a wave guide; both wave guide and electromagnet are in the possession of the proposed experimenters and no funds are being requested for their purchase. The manner in which the apparatus is to be used has been discussed above.

PARTS LIST – EDITORIAL COMMENT

Ed note: **The reader assumes all risk for costs, liabilities, or damages of any kind.** Please read the DISCLAIMER (see *Chapter 2*, p. 5) before committing time or money to the program. While the results of the 1994 experiment (see *Part IV*, p. 129) were encouraging, there can be no guarantee that the reader will have the same results.

The following list was provided in the original grant proposal. I have updated the prices by researching the cost of the same or equivalent equipment on eBay and other outlets.

The cost of the apparatus and the difficulty of assembling it are a barrier to repeating the 1994 experiment, but the height of the barrier depends on the qualifications and resources of the experimenter. The author strongly encourages the reader to enlist the aid of an electrical engineer who

---

[49] Ref. 6, p. 126

has theoretical and hands-on familiarity with microwave technology and understands electron paramagnetic resonance and dynamic nuclear orientation. If this requirement is met, the experimenters might be able to improve on the configuration suggested here and in the experimental report.

In 1994, dynamic nuclear orientation was achieved with the aid of an electron paramagnetic resonance (EPR) unit. An EPR unit is high-end equipment found only in university physics and chemistry labs. The price of an EPR unit puts it well beyond the reach of most amateur experimenters (quotes in the range of $250,000 were found). Having a "plug-and-play," professionally built EPR device eliminated a host of problems and delays that may arise in a homebuilt device.

In principle, however, dynamic nuclear orientation is not difficult to achieve. The parts list below allows construction of a unit that should theoretically yield the same results as an EPR machine. However, building the unit from scratch exposes the experimenter to problems that the EPR unit manufacturer eliminated in the course of research and development. These problems are probably not documented anywhere, which is why assembling and testing the device will require expert advice, preferably from an electrical engineer with experience in microwave technology. This cannot be emphasized enough.

The sample is placed in a waveguide and the waveguide is immersed in a powerful electromagnetic field. The electromagnetic field inside the waveguide must be strong enough to cause the electrons in the metallic sample to precess. The weight of the sample must also register on the scale in real time. Again, arranging these elements requires expert guidance from a good electrical engineer, as well as fine-tuned equipment.

If the equipment is faulty or it is arranged improperly, the experimenter will lose their investment and their "failure" report will not represent a valid test of the hypothesis.

The following list reflects minimum requirements for apparatus to conduct the proposed experiments, along with estimated costs. Low and high estimates are given based on price variability discovered while researching the list. The examples cited are not the only alternatives for each component.

REQUIRED COMPONENTS AND ESTIMATED COST

1) Lock-in amplifier: $1,000 – $5,500

A *lock-in amplifier* extracts a signal with a known carrier wave from an extremely noisy environment. Examples with prices:

a)   SRS Stanford Research Systems SR510 Lock-in Amplifier Unit Module (used), $1,000

b)   EG&G 7260 DSP (Digital Signal Processing) Lock-in Amplifier (used), $3,000

c) Perkin-Elmer Instruments 7265 DSP Lock-in Amplifier (used), $5,500

2) Microwave isolators (two): $450 – $2,000

A *microwave isolator* is a two-port device that transmits microwave power in one direction only. It is used to shield equipment on its input side from the effects of conditions on its output side; for example, to prevent a microwave source being detuned by a mismatched load. Examples below:

a) Farion TWG103D2-1 (used), $225 each

b) High-Power Varian Isolator (used), $350 each

c) FBI-28-SSESO Millitech Isolator (used; new over $1,000), $480 each

d) M/A-Com, Inc., R365S 26.5 to 40Ghz HP Isolator (used; new over $1,000), $480 each

3) One-watt traveling-wave tube amplifier: $130 – $4,000

A *traveling-wave tube* is a specialized vacuum tube that is used to amplify radio frequency signals in the microwave range. The TWT belongs to a category of linear beam tubes, such as the Klystron, in which the radio wave is amplified by absorbing power from an electron beam as it passes down the tube. Expert advice is required before purchasing, as indicated by the wide range in prices.

a) HP 491C Traveling Wave Tube Amplifier (used), $130

b) Hughes 8010H/8010H09F000 TWT/TWTA Traveling Wave Tube Amplifier 10W GHz 1-2 GHz (used), $4,000

4) Function generator: approx. $120

A *function generator* is electronic equipment or software used to generate different types of electrical waveforms over a wide range of frequencies. One of the common forms generated is the square wave used in the experiment. Integrated circuits (ICs) may also be used to generate waveforms.

HP 3310A (used), $120

5) Oscilloscope: $150 – $270

An *oscilloscope* shows constantly varying signal voltages as a two-dimensional plot of one or more signals as a function of time. Oscilloscopes are used to observe the change of an electrical signal over time, with voltage and time describing a shape that is continuously graphed against a calibrated scale. The waveform can be analyzed for properties such as amplitude, frequency, rise time, time interval, distortion, and more. Digital oscilloscopes may calculate and display these properties directly, rather than manually measuring the waveform versus the scales depicted on the instrument screen.

An oscilloscope can be adjusted so that repetitive signals can be observed as a continuous shape on the screen. A *storage oscilloscope* allows single events to be captured by the instrument and displayed for a relatively long time, allowing observation of events too fast to be directly perceptible.

a) HP 1740A Dual-Trace 100MHz 2-Channel Oscilloscope (analog, used), $150

b) DSO5102P Hantek Digital Oscilloscope 100MHz 2-Channel 7" WVGA DHL (used), $270

6) Directional coupler: approx. $50

A *directional coupler* is a passive device used in radio technology. It couples a defined amount of electromagnetic power in a transmission line to a port, enabling the signal to be used in another circuit. A directional coupler only couples power flowing in one direction.

HP Agilent 774D (used), $50

7) Microwave crystal detector: $160 – $300

Used to rectify an RF signal.

HP Agilent 423B Crystal Detector (used), $160 – $300

8) Low-noise amplifier: $125 – $500

A low-noise amplifier amplifies a low power signal without amplifying the accompanying noise, improving the signal-to-noise ratio.

9) Electromagnet (3000+ Gauss): $100 – $3,000

The magnet must be as powerful as a magnet found in a low-end MRI device. Military surplus electromagnets vary in cost; eBay is also a possible supplier. The electromagnet requires a DC power supply (see below).

10) DC power supply: $100 – $900

Changes alternating current to direct current to power electromagnet.

HP 6264B (used), $100

11) Gauss meter: $1,200 – $2,300

Detects and measures magnetic fields.

a) Lakeshore Cryotronics Model 410-HCAT Handheld Gaussmeter, $1,207

b) Pacific Scientific 5180 Gaussmeter, $2,300

12) Metallic sample: $1,000

This is a job for a custom metal shop. Pure Al_{27} (99.999%) is an absolute requirement. The sample is dusted with iron or chromium particles or inclusions as described in the experimental report. It is suggested that the sample should be heavier than the sample used in the experiment. The limit of the suggested scale is 100 grams. Again, a good electrical engineer may be able to devise a test chamber that can accommodate a much larger sample.

13) Weight scale: $250 – $2500

The scale requires software to track weight changes in real-time increments of milliseconds or, better, microseconds and feed the data to a computer. The software needs to match input from the microwave generator.

LW Measurements HRB103 (new), $250 (digital, 100 grams, .001 gram sensitivity; computer output available, but requires programming by technician not included)

14) Miscellaneous cabling: $500

15) Secure laboratory: $0 – $10,000

The price of a lab varies wildly, depending on whether or not the experimenter already has access. However, a secure laboratory is highly desirable, given the value of the equipment and the possible spying or theft of lab notebooks, data, and IP, as occurred at Kitty Hawk. *And the stakes here are much, much higher than Kitty Hawk.* If the experimenter is using a garage or a backyard workshop, it should have locks, alarms, motion detectors, and remotely accessible video monitors that feed recordings to the cloud. All files should be encrypted and backed up at a secure remote location.

16) Personnel: variable, from $0 – $70,000 each

The cost depends on whether the experimenter needs to hire an electrical engineer and someone to assemble equipment or not. The labor cost in 1994, as far as I know, was zero.

TOTAL ESTIMATED COST

Sum of low estimates: $5,335

Sum of high estimates: $102,940

[*Ed note: The above is the estimated price of a do-it-yourself experiment in which the experimenter builds the dynamic nuclear orientation apparatus. The cost of the experiment varies widely, depending on the quality of equipment purchased, and access to facilities and expert advice. Ideally, the experimenter knows microwave technology and works in a laboratory with easy access to an EPR. Access to an EPR improves the chances of success and eliminates most of the parts list.*

While the high and low estimates vary significantly, they do provide a ballpark figure that the experimenter can use as the starting point for a budget.]

SCHEDULE

Time to completion: Roughly six months

Purchase of equipment: Two months

Assembly of instrumentation: Two months

Experiments: Six weeks (includes time for correcting equipment errors and generating a final report)

If the experiments are successful, the project enters Phase 2: investor acquisition and research to determine engineering parameters for a propulsion system.

SUMMARY

This proposal requests funding for an experiment to test the predictions of the theory of Refs. 2, which, in 1960, advanced the assertion that, since a light signal (featured in Einstein's special theory of relativity) undergoes fluctuations in intensity, its apparent speed (as measured) may vary. This leads to a fundamental reformulation of the special theory of relativity and results in a unified field theory.[50]

If successful, this program will lead to extensive exploration and colonization of our planetary system and beyond, with considerably less cost than presently projected by space agencies using conventional rocket propulsion.

---

[50] See Refs 3, 4, 5, p. 126

REFERENCES (for 1989 grant proposal only)

1) M. A. Garbell, "Soviet Research on Gravitation, An Analysis of Published Literature," sponsored by Science and Technology Section, Air Information Division, AID Report 60-61, 379 pages. Distributed by U.S. Department of Commerce, Business and Defense Services Administration, Office of Technical Services, Washington, D.C., October 1960. [*Available online.*]

2) F. E. Alzofon, "The Origin of the Gravitational Field," *Advances in the Astronautical Sciences*, Vol. 5, Plenum Press, N.Y. (1960), a publication of The American Astronautical Society.

3) F. E. Alzofon, "Anti-Gravity with Present Technology: Implementation and Theoretical Foundation," AIAA/SAE/ASME 17th Joint Propulsion Conference, Colorado Springs, July 27-29, 1981, Paper Number AIAA-81-1608. [Available online at AIAA website; see p. 259.]

4) F. E. Alzofon, "The Unity of Nature and the Search for a Unified Field Theory," *Physics Essays*, *6* (1993) 599-608. [Reprints available through *Physics Essays*.]

5) F. E. Alzofon, "Light Signals, the Special Theory of Relativity and Reality," *Physics Essays*, *14* (2001) 144-148. [Reprints available through *Physics Essays*.]

6) C. D. Jeffries, *Dynamic Nuclear Orientation*, Interscience, John Wiley & Sons. N.Y. 1963

7) L. L. Foldy, "Elementary Particles and the Lamb-Retherford Line Shift," Phys. Rev. *93* (1954) 880.

8) H. E. Puthoff, "Gravity as a Zero-Point Fluctuation Force," Phys. Rev. A *39* (1989) 2333. [*Ed note: Dr. Puthoff and my father were colleagues at SRI in the late 1950s and remained in touch on a friendly basis through the 1990s or later. Their gravitation theories are similar in concept, but differ in significant respects.*]

9) G. E. Pake, "Fundamentals of Nuclear Magnetic Resonance Absorption," I. Am. J. Physics, *18* (1950) 438, and II, Am. J. Physics, *18* (1950) 473.

Part IV

The 1994 Experiment

Chapter 20

Introduction

by David Alzofon

THE RATIONALE for the experiments and the circumstances surrounding them have been detailed elsewhere (see *Chapter 15*, p. 87, *Chapter 19*, p. 117). Here we will discuss *weaknesses* in the setup, not as a confession, but rather to explain why my father regarded the results as unpublishable. My father, who had considerable experience with setting up and running experiments, did not believe any of the problems about to be discussed undermined the conclusion that the gravity control effect had been observed. It is hoped that pointing out these blemishes will enable the next experimental team to improve on the original design and generate a publishable paper.

First a few words about goals. It was never his intention to mount a foolproof, perfect experiment. The funding simply wasn't available to buy equipment that would allow that. Rather, he wanted to see if the gravitational effect would manifest at room temperature with the chosen materials. Since all textbook experiments on dynamic nuclear orientation had been conducted at temperatures close to absolute zero, this was something new. One of his predictions was that it would be possible to sustain nuclear orientation at room temperature using aluminum. If it worked, it would be especially fortunate for spacecraft construction, because aluminum is strong, cheap, and abundant.

As he described the experiment to me, he was seeking a qualitative, not a quantitative verification of the hypothesis. In particular, he predicted that the sample weight would spike during the "driving phase" of nuclear orientation when the microwave field was on, then drop as soon as the microwave field was turned off. Each microwave pulse would cause more orientation to accumulate in the sample, hence more weight to be lost each time the pulse was turned off, somewhat in the manner that pushing a swing increases the arc. This pattern was in fact observed, as can be seen in chart AF2001 (on the cover and below), and elsewhere.

Absence of the spikes from the pattern observed in the first trial (Plot AF0001, p. 142) is probably why he wrote: "No correlation between weight alteration and microwave field alteration was observed." I suspect what he meant to say was that weight loss *was* observed, but the spikes were not, so it was impossible to correlate the microwave pulses with the weight loss. As I look at Plot AF0001, it appears to me that there *are* two spikes, but the second spike is small enough to be equivocal. I must defer to FA's judgment and provide an accurate transcription of the results as he saw them.

One of the difficulties I had in interpreting the charts was in determining the meaning of the numbers on the vertical axes. As he put it in the introduction (p. 139):

> *The grams per volt factor is calculated from the nominal gauge factor of the strain gauge used in the weight alteration measurement (0.001V per gram), and the amplification gain. For example, if the amplification gain is 3000, this factor is calculated to be 1/0.001 x 3000 = 0.333333 grams per volt.*

> *Assuming the preceding factors are essentially correct, we obtain a total conversion factor 9.53764E-7 x 0.333333 = 3.17891E-7 gm weight per total counts. The latter value was then multiplied by the total number of counts to obtain incremental weight changes.*

While this explanation might have been helpful to someone who was already familiar with its meaning or well-versed in the interpretation of data from a scale setup like the one in the experiment, it was opaque to me. I showed the explanatory text and Plot AF0002 to three sophisticated readers, one of them an electronics genius and computer science instructor at a college in northern California, and none of them could figure it out completely. The college instructor (Master's degree and PhD candidate in mathematics at Stanford) wrote back the following:

> *For the vertical on the chart [AF0002, p. 143], I think that "- E - 4" means "times 10 with an exponent of -4", or, in other words, "times .0001". "- E - 5" would mean "times .00001".*

> *Therefore the numbers are (bottom to top)*

> *-1.4 x .0001[or -.00014],*

> *-1.2 x .0001[or -.00012],*

> *-1.0 x .0001[or -.00010],*

> *-8.0 x .00001[or -.00008] etc.*

> *The number at the top, 7.3 x .000000000001, doesn't appear to be in sequence with the others. I don't know anything about the physics, and I have no idea what a "volt/gram" could mean.*

More about the meaning of the vertical axis, I can't say, except that I suspect that the markings on the axis all show the same increments of weight, but that the data associated with each increment changes according to readouts from the computer. Admittedly, this is a guess.

It was disappointing to discover that the sample weighed only 1.1143 grams. The Mettler scale, however, was very sensitive and designed to show weight changes on the order of 0.01 *mg*. In other words, a very small change in weight would register large on the equipment. Suggested improvement:

a much heavier (more expensive) sample, which would require a larger test cavity.

Another difficulty was the impossibility of precisely correlating the microwave pulses with the weight changes. An improved version of the experiment would attach data from the microwave field generator to the same clock as data from the scale. The experiments do show peaks and valleys in the right places on the timeline. What can't be told for certain is the correlation between the peaks and the square-wave pulses. My father remarked on this after I signed the NDA. The correlation would undoubtedly provide useful data. Without it, the results are unpublishable.

Because of a problem with the computer system (see p. 141), the tests were limited to two microwave pulses. This problem could easily be eliminated with better equipment.

Could the weight alteration be caused by anything other than the configuration of fields? The material of the sample was paramagnetic, so magnetic fields alone would not affect it. The point is moot anyway, because the magnetic field surrounding the test cavity originated in the EPR and was constant, with no reversal in polarity. Constancy in the field was an absolute requirement for the experiment. Microwaves alone would have no effect on weight, either.

One physicist said, "How do I know this isn't just floor vibration?" when he saw the charts. Floor vibration *is*, in fact, visible on the charts in the ultra-fine-toothed shimmying along the length of each line, such as can be seen in chart AF0001 (p. 142). The test chamber was isolated and vibration in the floor or the equipment did not register appreciably on the scale, so vibration is not a plausible explanation for a *weight* change. This did not impress the physicist, however, who seemed anxious to dismiss the results. This kind of reaction is part of the reason nothing was submitted to a journal. Standard controls, such as running the test with the microwave field but without the constant magnetic field and vice versa, were conducted. In each case the weight alteration vanished.

Some readers may be disappointed that there are no pictures of an object floating. After all, isn't that the point? YouTube is replete with demonstrations of "antigravity devices" making things float, so what's so great about removing 80% of the weight from a one-gram chunk of aluminum? Such desires are understandable, but more appropriate to a movie script. The goal of the experiment was to prove that the gravity control effect occurs at room temperature in the pattern and the amount *predicted by theory*. In order to observe weight changes, it was necessary to attach the metallic sample to a scale and obtain continuous readouts in real time.

Weight changes *did* occur with the predicted pattern, which has enormous implications. It means Phase I of the three-phase program—proof of concept—was complete. Phase II would begin by doing more tests, gathering more experimental data, and determining how to transfer the effect to the hull of a craft in the most efficient manner. Next, a drone GCV would be built and test flown. Data gathered from the drone test flights would be used to launch Phase III, construction of a manned space vehicle and any number of terrestrial flying machines, such as the one described in *Chapter 5, A Trillion-Dollar Technology* (p. 19). Making a metal ball bounce around in an "antigravity chamber" might have been possible, but it would have provided little more than a "feel good moment," and nothing in the way of useful data. The carefully timed and measured perturbations in the weight of a one-gram sample were far more significant.

One might even say they were earthshaking.

Chapter 21

Gravity Control – Preliminary Experiments

By Dr. Frederick Alzofon

Dates of Experiments: May 26, June 17, and June 18, 1994

INTRODUCTION

The following report is an account of preliminary experiments designed to establish the feasibility of weight alteration by control of the orientation of atomic nuclei.

The theoretical justification of the process used in these "proof-of-concept" experiments has been described in two papers: "The Unity of Nature and the Search for a Unified Field Theory," by Dr. F. Alzofon (*Physics Essays*, Volume 6, pp. 599-608, 1994), and "Antigravity with Present Technology – Implementation and Theoretical Foundation" (proceedings of 17th Joint Propulsion Conference, Colorado Springs, CO, 1981, AIAA document reference no. AIAA-81-1608).

Briefly characterized, the theory predicts a correlation between rapid alterations in the orientation and the randomization of atomic nuclei with a net magnetic moment, and the weight of the specimen containing the nuclei. The theory also predicts a pattern in the weight alteration response (see p. 136 below).

The experiments described in this report were designed to demonstrate this correlation and the feasibility of generating the predicted effect. The experiments were completely successful in this goal. Owing to the limited nature of the correlations sought, it was not necessary to perform highly accurate measurements.

Selection of the atoms whose nuclei were to be oriented was determined by the desire to maximize the effect desired. For this reason, metallic aluminum was chosen, as the nuclei of aluminum have a large magnetic moment.

The orientation of the aluminum nuclei was accomplished by means of dynamic nuclear orientation, although other means of orientation exist. Dynamic nuclear orientation is a well-known process that leverages the precisely timed flipping of large numbers of electrons to orient atomic nuclei. It is primarily used for assaying organic molecules. Devices manufactured by JEOL sell for roughly a quarter of a million dollars, but the experiment demonstrated that the effect can be achieved much more cheaply. (Dynamic nuclear orientation is not to be confused with nuclear magnetic resonance; they are two distinct processes.)

The following is a step-by-step breakdown of the method used to achieve dynamic nuclear orientation in the experiment:

1) A specimen of metallic aluminum was placed in a constant magnetic field (about 3300 gauss) at room temperature. The magnetic field causes the electrons in the metal to precess at the Larmor frequency (about 9.5 GHz).

2) A pulsed microwave field was then applied to the specimen. The field has the same frequency as the Larmor frequency of the metal electrons and has a magnetic field vector oriented at right angles to the constant magnetic field. The crossed fields cause the electrons to precess so violently that they tip over *en masse*. Their close proximity to the atomic nuclei causes the nuclei to also tip over and become oriented. The nuclear orientation can be thought of as a consequence of the preservation of angular momentum.

During the process of nuclear orientation we expected the weight of the specimen to change. The theory predicted a slight, transitory weight increase. In addition, when the microwave field is cut off, we expected the weight of the specimen to alter during the randomization of the nuclear magnetic moments. The theory predicted the weight would fall during this part of the cycle. Repeated cycling of the orientation and relaxation of nuclei would act like a pump, causing the weight loss to accumulate dramatically and eventually cascade toward complete weightlessness.

The rate at which the microwave field is pulsed depends on the relaxation time for oriented nuclei. At room temperature this is 6 ms; accordingly, 6 ms was the pulse duration chosen for most of the experiments. However, to demonstrate the critical character of this time interval, a different interval was chosen for some of the experimental trials. In the latter case, the desired correlation was not observed.

In summary, the predicted correlations between microwave field variations and weight variations were conclusively demonstrated, verifying the validity of the theoretical foundation for the experimental design, and at the same time establishing the viability of gravity-control technology for vehicle propulsion.

Chapter 22

Experiment 1, Setup

May 26, 1994

INTRODUCTION

The tests conducted in Experiment 1 were of a preliminary nature; they functioned as a "get-a-hands-on-feel" for use of the equipment. The tests were conducted on May 26, 1994, beginning at 0930. Personnel present were Dr. Frederick Alzofon, [*name unknown; withheld by F.A.*], and the owner operator of the equipment, [*name unknown; withheld by F.A.*].[51]

Eight tests were conducted. The computer-generated data from these appears on charts AF0001 through AF0008 (pp. 142 – 149). An additional test was run on a sample of pure aluminum without iron inclusions, and the weight alteration effect was observed here as well.

The equipment used consisted of an electromagnet producing a constant (dc) magnetic field, and an x-band microwave system and interfacing systems. The specimen whose weight was measured was placed within a resonant cavity in the dc magnetic field; the microwave field generated within the cavity was pulsed to produce the observed weight variation of the sample, measured by a sensitive transducer.

The sample on which measurements were made was composed of an aluminum-iron (Al-Fe) alloy with 97.5% Al by weight (99.999% pure) and 2.5% Fe by weight (99.99% pure). The alloy was powdered; the powder-particles' sizes were less than one micron.

The metallic powder, weighing 0.5047+/-.01mg gm, was mixed with a casting plastic and formed into a cylindrical shape having a diameter of 0.204+/-.001 inch and a length of 0.752+/-.001 inch. The total weight of the resultant sample was 1.1143+/-.01mg gm.

---

[51] The photocopies of the report given to me by my father had only blank spaces where the names had been. Evidently he wanted to protect the identities of the other participants. I never learned their names.

Since the manufacturer of the plastic had no technical data about any effects the plastic might experience or cause in an excited microwave cavity, we performed tests to determine what such effects might be. No effects were observed to be caused by the plastic and no physical alteration in the plastic was observed.

MECHANICAL ADAPTOR APPARATUS

A plastic housing containing the sample and fitting into the microwave resonant cavity was constructed from a microwave plastic manufactured by the 3M Company. An aluminum adaptor cylinder to which the plastic housing is fastened connects the housing to the microwave resonant cavity. The test sample is connected to the weighing transducer mounted on the aluminum cylinder by means of a small-diameter plastic rod weighing 0.1162 gm.

MICROWAVE SYSTEM CALIBRATION

The microwave system was calibrated by a former owner who had over eight years of experience building, debugging and calibrating equipment. The procedure for calibration is complex and may be available in a manual which is, at present, not locatable. The following simplified calibration procedure was used for the present experiment:

a) The empty resonant cavity's resonant frequency was measured to be 9.550 GHz.

b) The microwave plastic housing was inserted into the resonant cavity and a resonant frequency of 9.545 GHz was observed, a reduction of 0.005 GHz from the empty cavity value.

c) The casting plastic was inserted into the cavity and a resonant frequency of 9.530 GHz was observed, somewhat more of a reduction than in b) above for the plastic, but still small enough to be considered negligible.

d) The Al-Fe cylindrical sample was inserted into the cavity and a resonant frequency of 9.235 GHz was observed, a difference from the unperturbed value of about 3.3 percent. This difference was also considered negligible.

The RF power input to the cavity was measured to have the value of 0.125 +/-0.005 watts with a Q of 1541 +/- 10.

The dc magnetic field of the electron paramagnetic resonance unit was measured to be 3008 +/- 5 gauss.

With reference to the timelines exhibited in the plots for experiments AF0001 through AF0008, and experiments AF2001 through AF2006, respectively, there appears to be a leading 5 – 6 ms off-time for the microwave field, beginning from time t = 0, for the first set of experiments.

In contrast, for the second set of experiments, there is no off-time. In general, the data for the set of experiments AF2001 through AF2006 appears to be much clearer and more accurate than the data for experiments AF0001 through AF0008.

SCALE

The weighing scale used to weigh the test samples was a Mettler high-accuracy scale capable of 0.01 mg weight-change resolution.

DATA SET DERIVATION

Each of the data sets consists of 1025 x-y pairs; the x-value represents a value of the time interval, and the y-value represents weight-base data after processing as described below.

The x data is sampled at 50 microsecond intervals, and since there are 1024 x-values, the total length of the time interval is 50 x 1024 = 51.2 milliseconds (ms). A "scan" is defined as the collection of 1024 data points, sampled at equal time intervals (50 microseconds), totaling a time interval of 51.2 ms.

Corresponding to each value of x (a "time bin") there is stored a value of weight data which has been averaged over the number of scans. The data is stored as a six-bit word chosen from a possible 64 discrete levels for each scan; after n scans, $64n$ values of the weight data have been averaged.

The following notation with respect to the BioMation computer is used: "1024 scans/BS = 32" means that 32 groups of 1024 x-y pairs were averaged.

SCALING CALCULATIONS

The data acquisition equipment used for the experiments provided data in the form of dimensionless numbers or counts in a range between zero and a positive integer determined by the number of scans and input range setting. The number of counts is converted to increments of weight gain or loss by multiplication of two factors: volts per total counts and grams weight per volt.

The factor with dimensions of volts per total counts is calculated by the product: input range (0.5V for all the experiments) in volts, divided by the number of scans and by the number of digitization levels (64). For example if 8192 scans were averaged, the conversion factor would be 0.5 / 64 x 8192 = 9.5367E-7 volts /total counts.

The grams per volt factor is calculated from the nominal gauge factor of the strain gauge used in the weight alteration measurement (0.001V per gram), and the amplification gain. For example, if the amplification gain is 3000, this factor is calculated to be 1/0.001 x 3000 = 0.333333 grams per volt.

Assuming the preceding factors are essentially correct, we obtain a total conversion factor 9.53764E-7 x 0.333333 = 3.17891E-7 gm weight per total counts. The latter value was then multiplied by the total number of counts to obtain incremental weight changes.

Note: The values of gauge factor, amplifier gain, and analog-to-digital conversion accuracy were not known to the accuracy desired and were only estimated.

Chapter 23

Experiment 1 – Tests and Results

May 26, 1994

OVERVIEW

Nine tests were conducted. After several tests of multiple-period microwave input pulses, it was found that the microwave test equipment computer system would lock-up on tests comprising more than two periods; it was therefore decided to limit the test to two pulses, as sketched below. In accord with a suggestion by F. Alzofon, the on-off time intervals for the pulses were selected to be .008 seconds each. It was thought, for reasons which do not appear clear at this time, that the lifetime for polarized aluminum nuclei to relax at room temperature might be longer than the .006 sec reported in the literature.[52]

Microwave field	OFF 1 ms	ON 8 ms	OFF 8 ms	ON 8 ms	
ms count	0	1	9	17	25

---

[52] J. J. Spokas and C. P. Slichter, "Nuclear Relaxation in Aluminum," *Physical Review* 113 (1462) 1959

EXP. 1, TEST 1 - PLOT AF0001

Result: No correlation between weight alteration and microwave field alteration was observed [see comment, p. 131]. In processing the data, 32 scans of 1024-point data sets were averaged.

TEST 1 – DATA

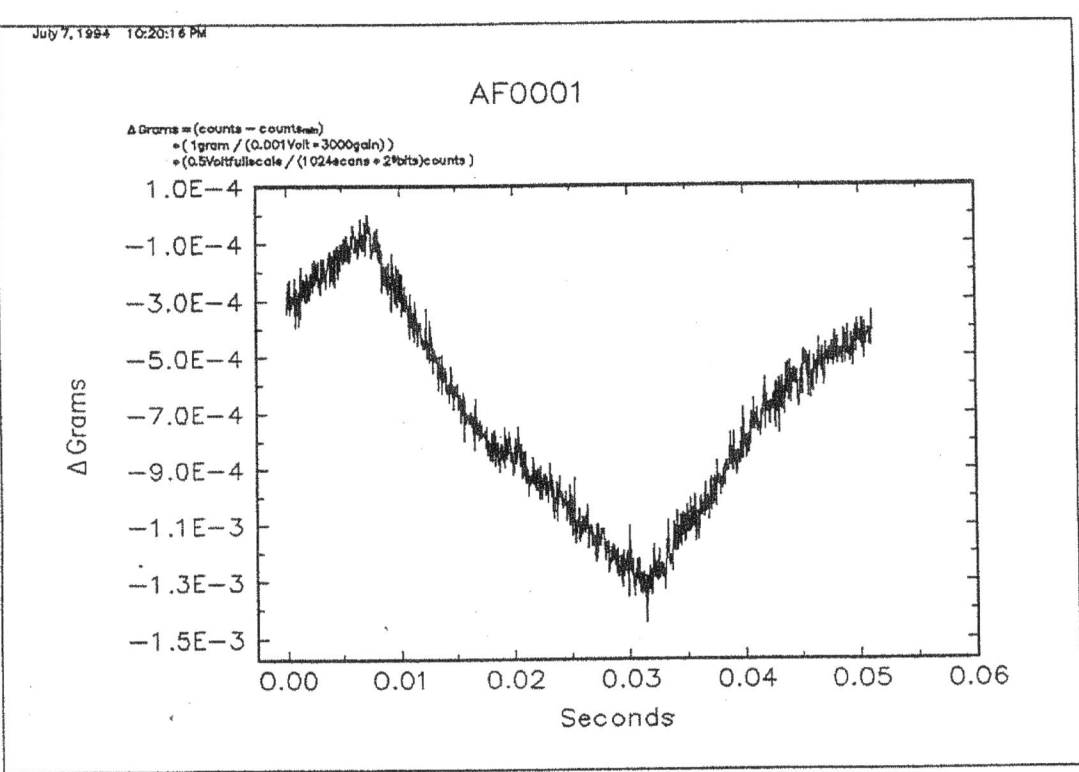

Ed notes:

Parts of the legend above the chart is illegible. As best as can be determined, it reads as follows:

$\Delta\ Grams$ = *counts* – *counts*$_{min}$

= 1 gram / (0.001 volt – 3000gain)

= 0.5 voltfullscale/(1024scans – 2 bits) counts. [The superscript between "2" and "bits" is illegible.]

Note: All plots had the same date stamp (July 7, 1994). Apparently the setting was not adjusted.

EXP. 1, TEST 2 – PLOT AF0002

For Test 2, the on-intervals were altered to 6 ms in accord with the value of the relaxation time for oriented aluminum nuclei in a metal at room temperature,[53] as illustrated below.

Microwave field	OFF 1 ms	ON 6 ms	OFF 6 ms	ON 6 ms	
ms count	0	1	7	13	23

Result: As exhibited in Plot AF0002, there is a well-defined correlation between weight variation and microwave field intensity variation. Note: In processing the data, 32 scans of 1024-point data sets were averaged.

TEST 2 – DATA

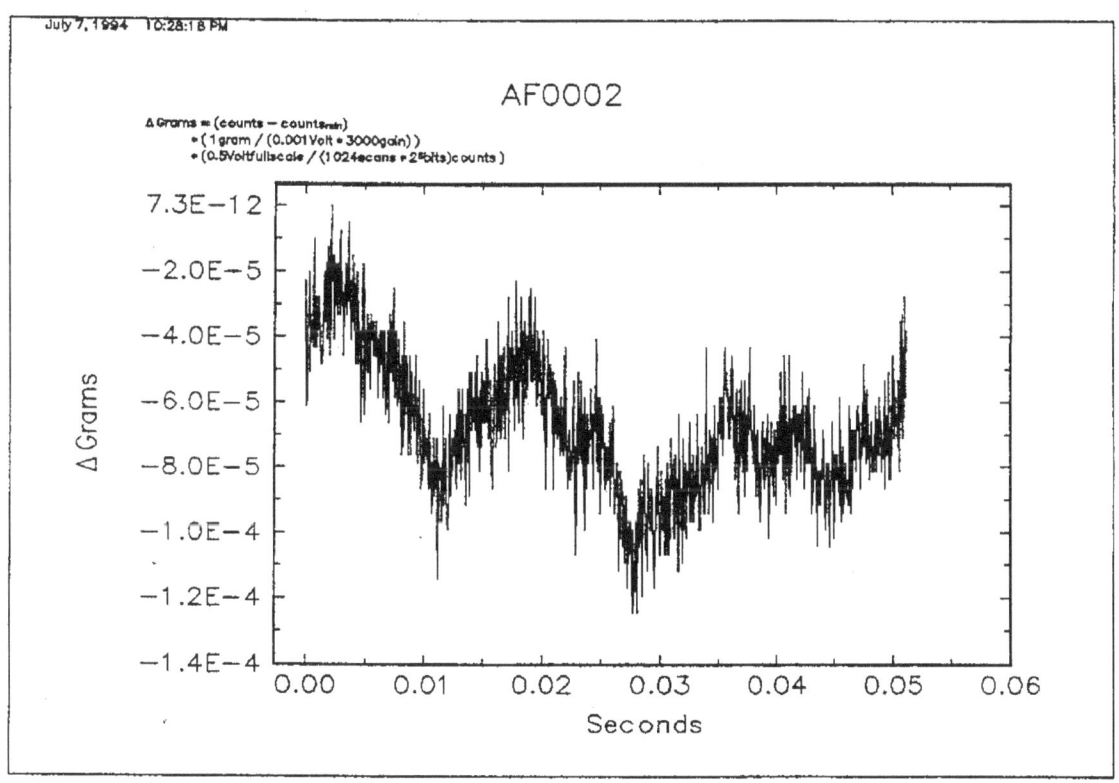

---

[53] Ibid, Spokas and Slichter, p. 297

143

EXP. 1, TEST 3 – PLOT AF0003

For Test 3, the conditions of Test 2 were repeated except that more data-point averaging was introduced.

Result: As exhibited in Plot AF0003, much of the noise shown in Plot AF0002 has been suppressed by the increased averaging, and the correlation between weight variation and microwave field intensity is more prominently displayed.

TEST 3 – DATA

EXP. 1, TEST 4 – PLOT AF0004

For Test 4, the on-off intervals were all altered to 5 ms, preceded by the 1 ms off-interval shown in the diagram for Test 2 (p. 143).

Result: The correlation of weight variation and microwave field variation is well-defined, as exhibited in Plot AF0004, but the noise level is somewhat larger than in Test 3.

TEST 4 – DATA

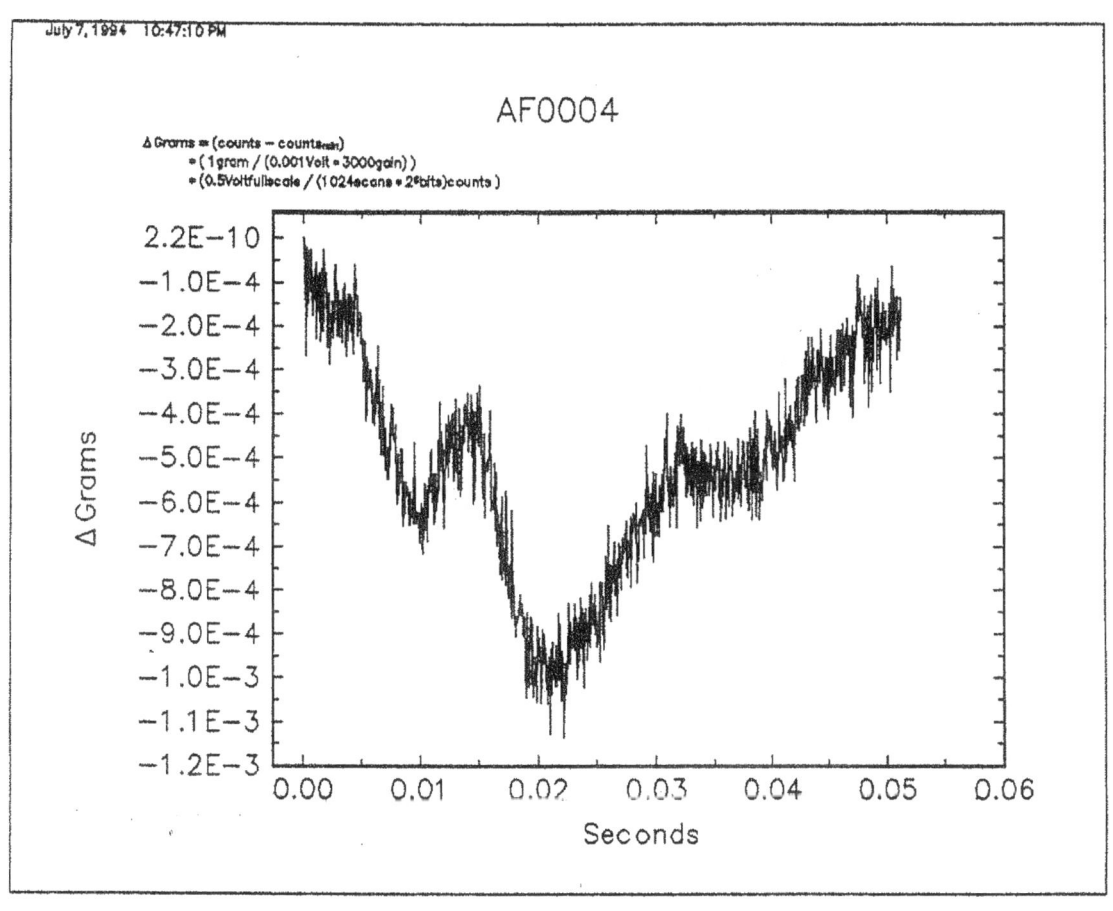

EXP. 1, TEST 5 – PLOT AF0005

For Test 5, the on-off intervals were all set equal to 3 ms, preceded by the 1 ms off-interval as shown in Test 2.

Result: The correlation between weight variation and microwave field variation was not clearly present, as shown in Plot AF0005.

TEST 5 – DATA

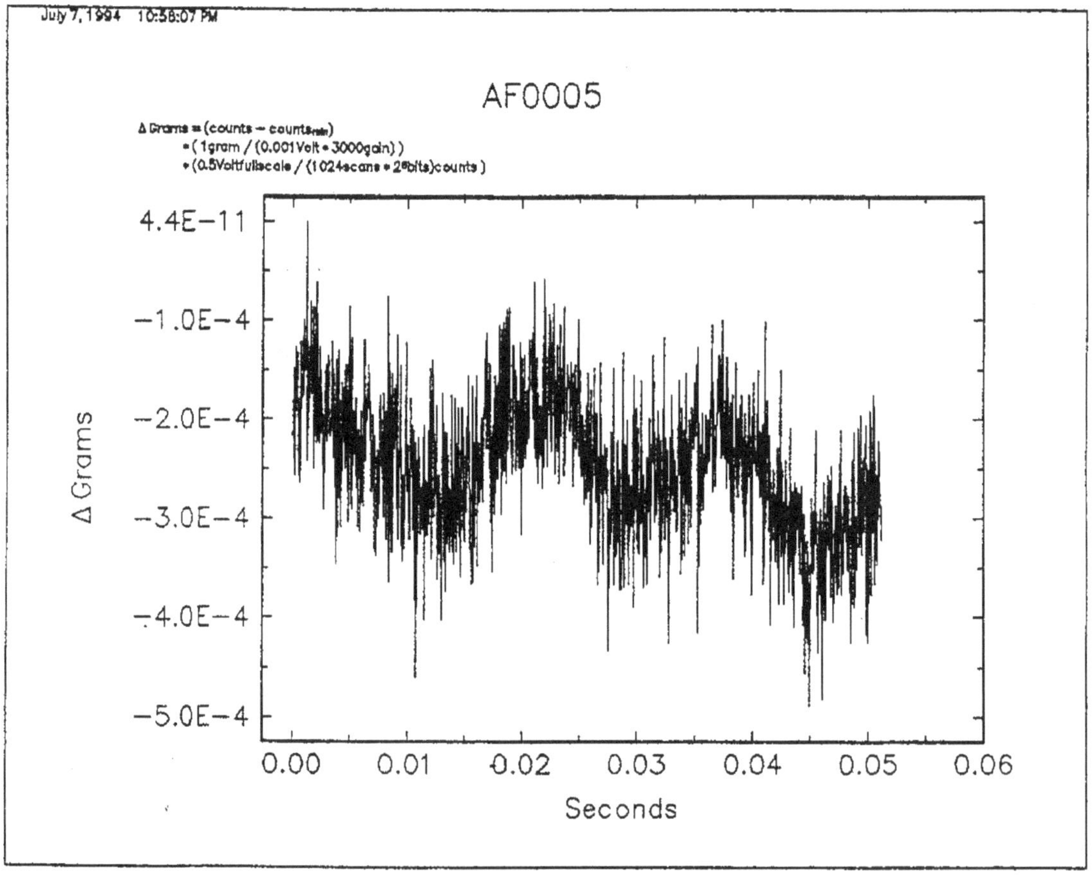

EXP. 1, TESTS 6, 7, AND 8

For tests 6, 7, and 8 some variations in test conditions were instituted. However, during this period changes were being made in the data-gathering instrumentation faster than could be followed by the experimenters and the data exhibited in Plots AF0006, AF0007, and AF0008 could not be interpreted readily in terms of well-defined experimental conditions.

EXP. 1, TEST 6 – PLOT AF0006

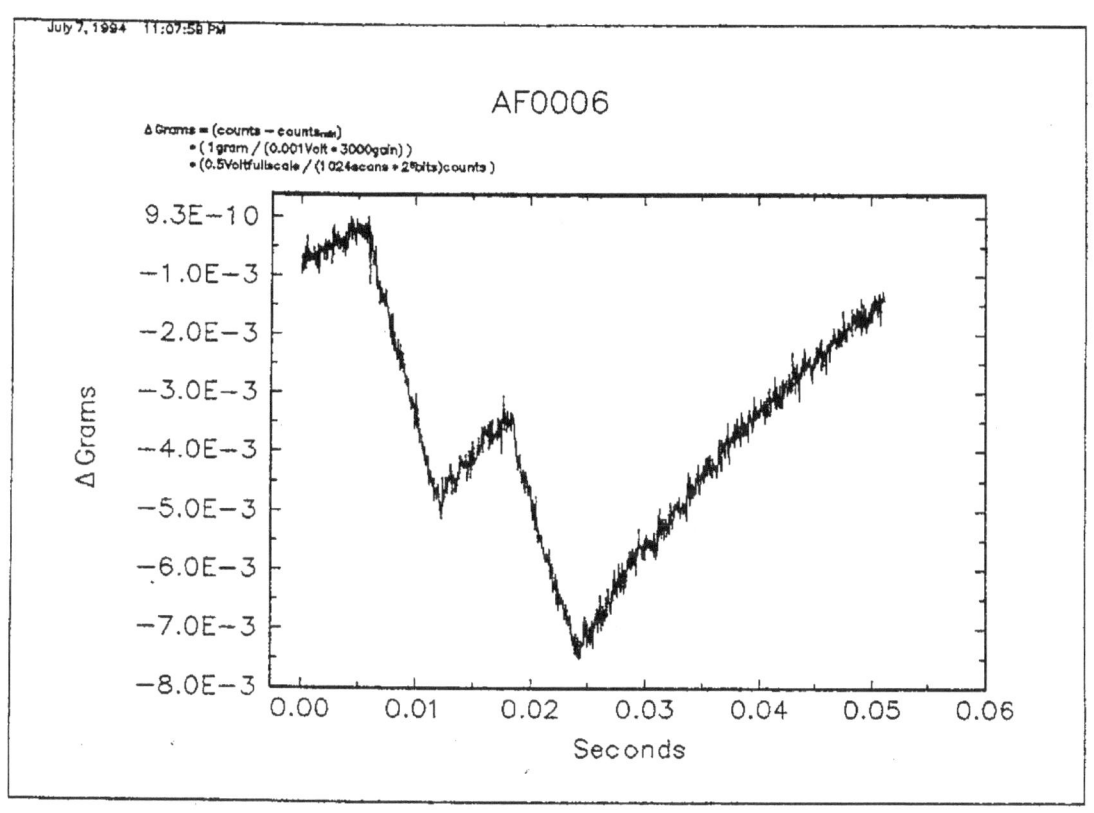

EXP. 1, TEST 7 – PLOT AF0007

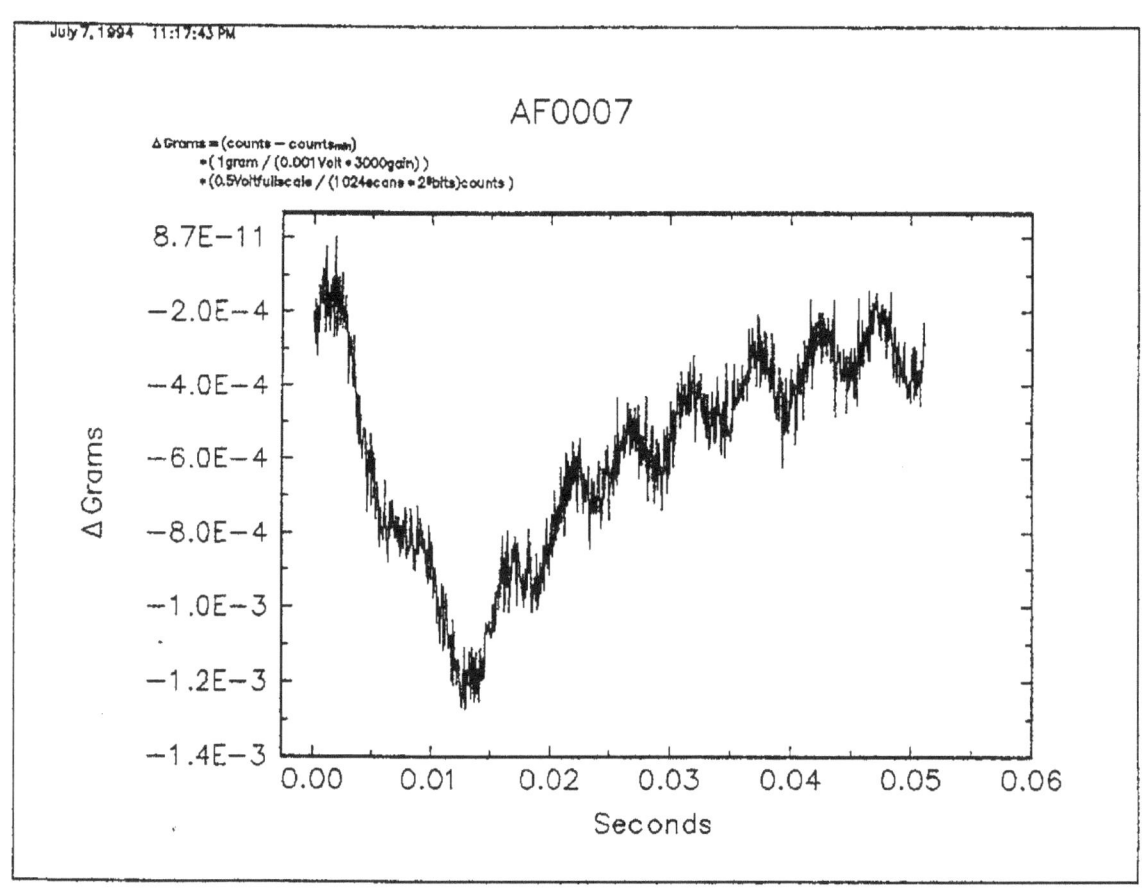

EXP. 1, TEST 8 – PLOT AF0008

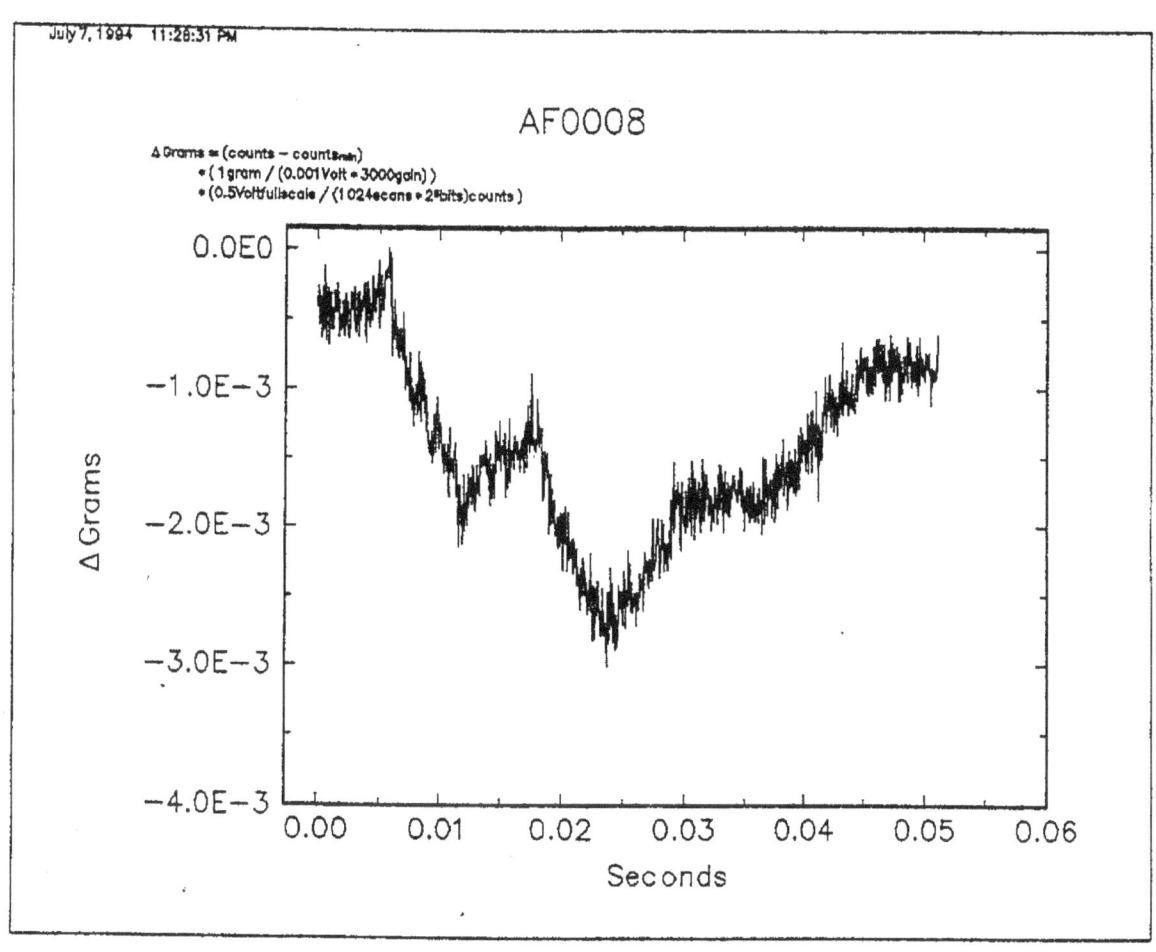

EXP. 1, TEST 9 – ALUMINUM WITHOUT IRON

An additional test was conducted using pure aluminum without iron inclusions. The microwave cycle was the same as in Test 3 above. A correlation was observed between weight variation and variation in microwave field intensity. The observation is in accord with data on the orientation of aluminum nuclei in metallic aluminum in *Physical Review*[54] and the model for a correlation between nuclei orientation and disorientation and weight alteration of a given sample.

No chart exists for this test run.

End of Experiment 1

The experiments were terminated at 1450 hrs.

---

[54] Ibid, Spokas and Slichter, p. 297

Chapter 24

Experiment 2

June 17, 1994

CALIBRATION AND TESTING FOR A JUNE 18, 1994, DEMONSTRATION

At 1736 hrs, June 17, 1994, calibration for a June 18, 1994, demonstration was begun. The demonstration is for a potential investor. The cylindrical sample used for the calibration and demonstration is composed of the same Al-Fe powdered material used in Experiment 1 (May 26).

The same apparatus used in Experiment 1 was employed in calibration and testing. The on-off intervals for the microwave input were the same as employed in Test 2 of Experiment 1 (i.e. 6 ms).

The correlation of weight and microwave field alteration was well-defined and corresponded to Plot AF0006, as measured with a scale on a computer screen. It was then decided that the apparatus was ready for the demonstration. AF0006.dat file was duplicated, copied to another file and labeled AF2001.dat for use in the June 18 demonstration, since the system has worked well with this data file, giving the best output for the 6 ms on-off intervals for the microwave field.

End of Experiment 2.

Chapter 25

Experiment 3

June 18, 1994

DEMONSTRATION FOR A POTENTIAL INVESTOR

At 1400 hrs the experimental procedures began with a check to verify that the June 17 calibration was still valid and the apparatus was functioning properly. All checks of the electronic test equipment were performed in the same manner as in Experiment 1 (May 26). It was found that all system parameters were functioning properly.

Present at the experiments were Frederick Alzofon, [*names withheld by F.A.*].

For Experiment 3, 8192 "time bins" were programmed and an average was taken over 256 values of weight increments for each time bin. The microwave input pulses were chosen as in Experiment 1, Test 2, unless otherwise stated.

Tests, results, and data charts are shown below.

EXP. 3, TEST 1 – PLOT AF2001

The results of the experiment for Test 1 are shown in Plot AF2001. The correlation between microwave intensity and weight alteration is very clear.

TEST 1 – DATA

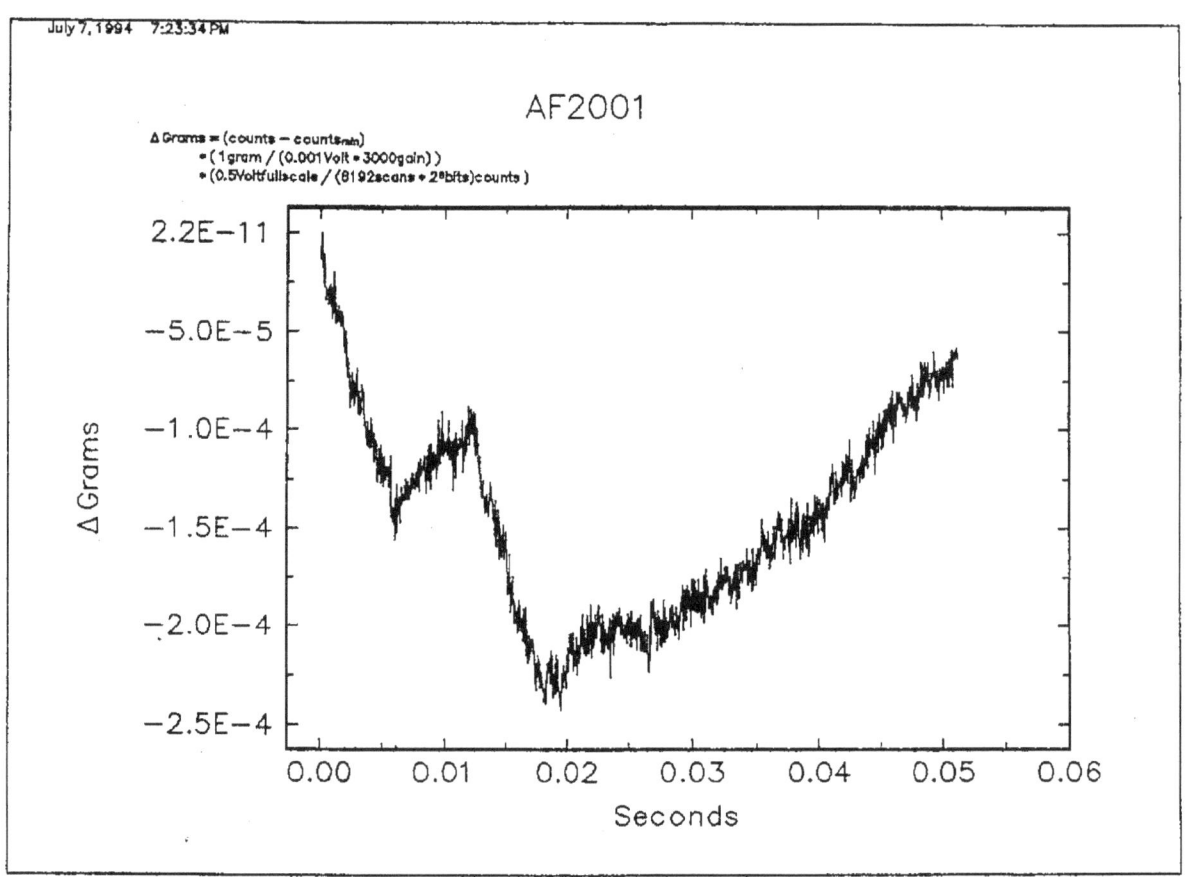

EXP. 3, TEST 2 – PLOT AF2002

The result of Test 2 is shown in Plot AF2002. Some correlation is evident between microwave intensity and weight alteration. There were some difficulties in computer processing of the data and these were reflected in the lack of a correlation, such as clearly shown in Test No. 1.

TEST 2 – DATA

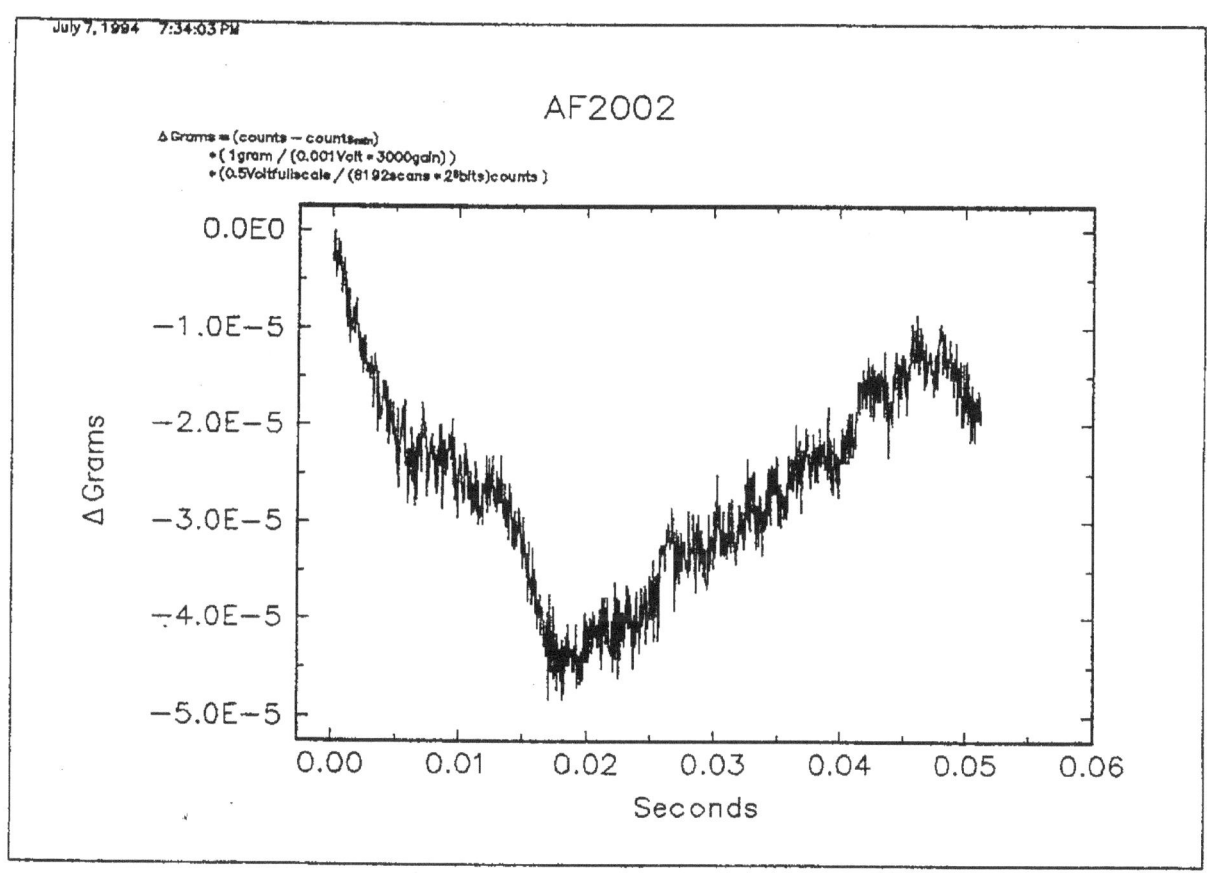

EXP. 3, TEST 3 – PLOT AF2003

Having corrected the data processing problems of Test 2, the results of Test 3 are depicted in Plot AF2003. The correlation between microwave field intensity and weight increments is very clear.

TEST 3 – DATA

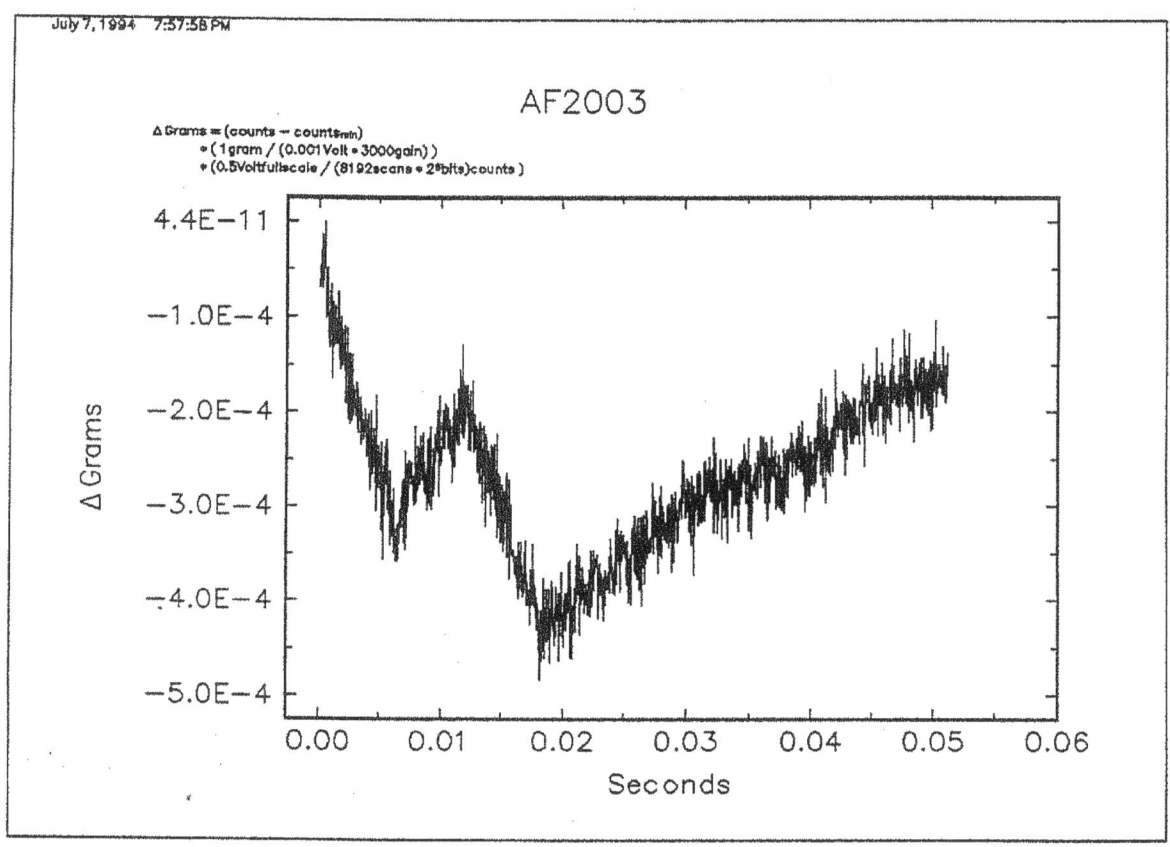

EXP. 3, TEST 4 – PLOT AF2004 (CONTROL)

The current supplying the electromagnet to produce a constant magnetic field was switched off for Test No. 4 in order to test the role of the constant magnetic field in the correlations noted above. There remained a weak residual magnetic field whose magnitude was not measured. The correlation between microwave field intensity and weight increments is still present.

It is felt that a plausible explanation for this persistence is that the fractional alteration in resonant microwave frequency is equal to the fractional variation in the constant magnetic field. Since the resonant frequency is so large (about 9.5 GHz), the bandwidth must also be very large, corresponding to alterations in the magnetic field. Plot AF2004 illustrates the correlation observed.

TEST 4 – DATA

EXP. 3, TEST 5 – PLOT AF2005 (CONTROL)

In Test 5, both the constant magnetic field and the microwave field were switched off. Plot AF2005 shows no variation in weight recorded.

TEST 5 – DATA

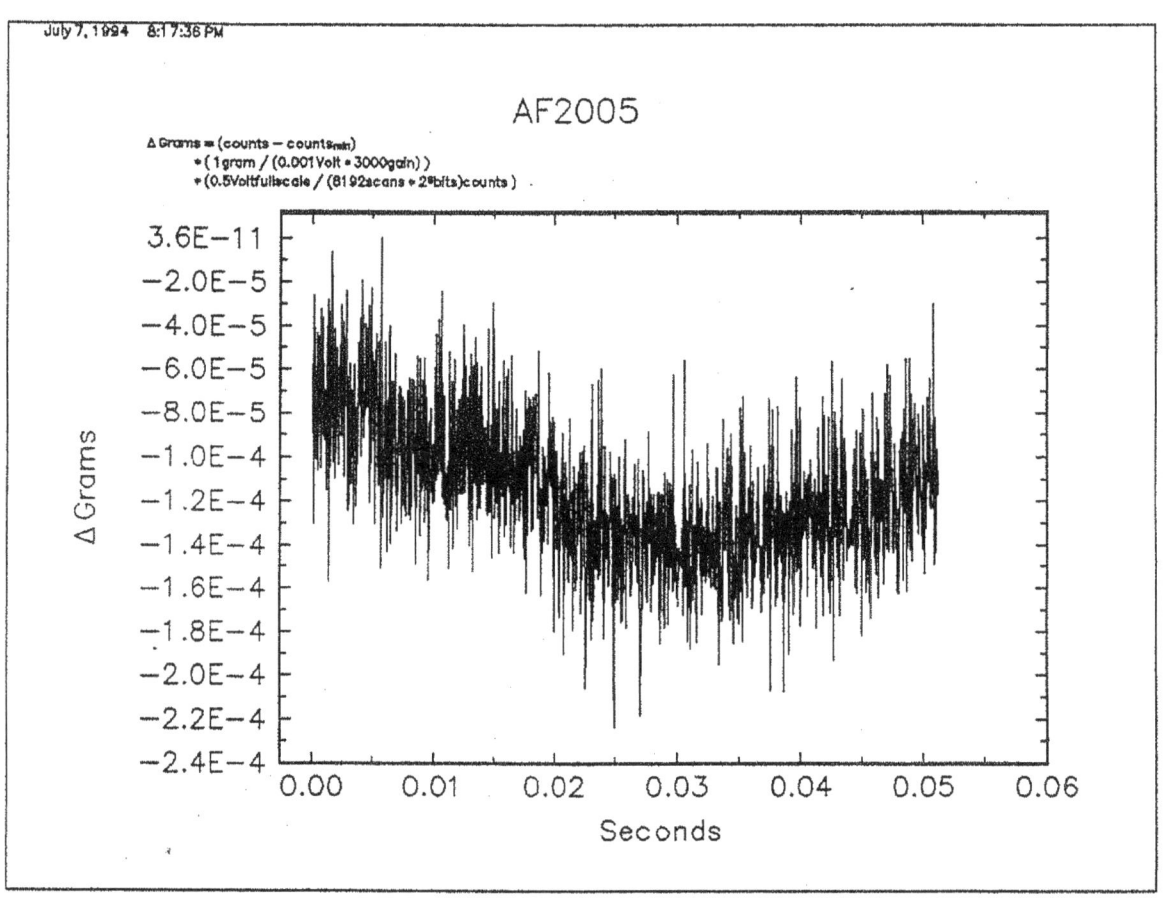

EXP. 3, TEST 6 – PLOT AF2006

In Test No. 6, both the constant magnetic field and microwave field were restored, as in Tests No. 1 to 3. As Plot AF2006 shows, the alteration of weight and the microwave field variation are correlated.

TEST 6 – DATA

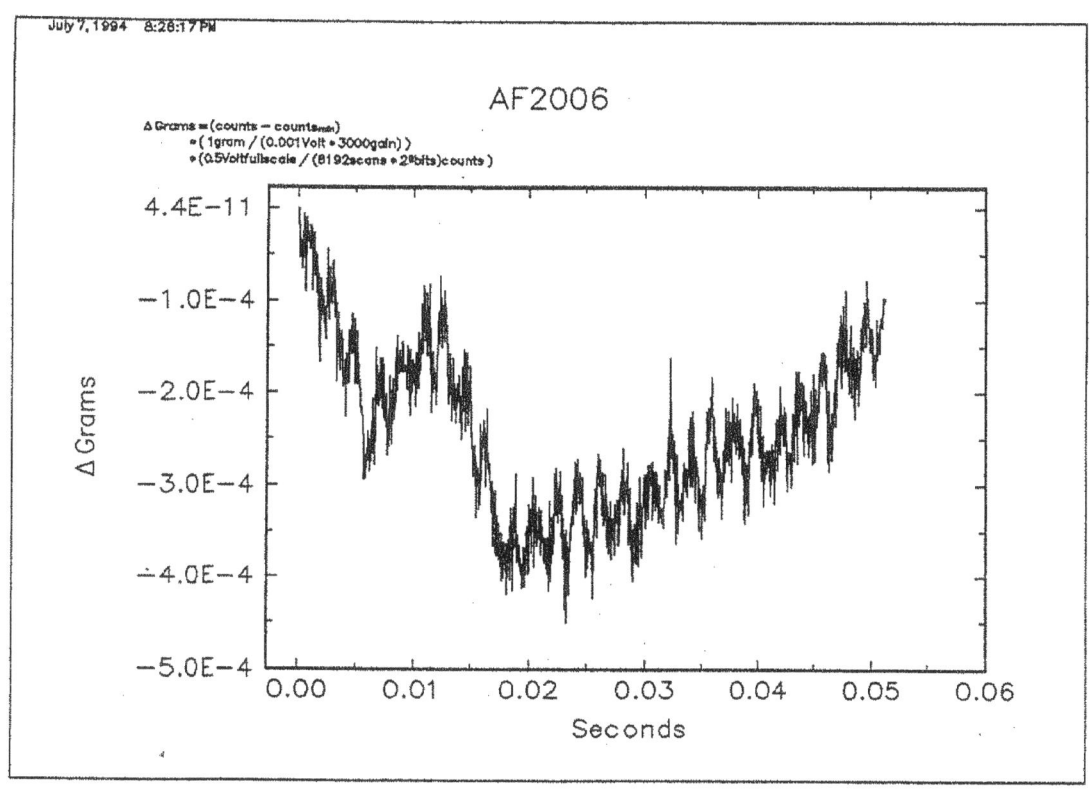

End of Experiment 3

CONCLUSION

The predicted correlation between pulsed dynamic nuclear orientation and weight alteration has been demonstrated. The weight fluctuations display the pattern predicted by the theory: an upward spike, followed by a drop in weight. The second drop, correlated with the second pulse, is more precipitous than the first, as predicted. When microwave pulses ceased after the second pulse, the weight returns to normal in a smooth line as nuclear orientation in the sample decays.

Chapter 26

What Next for Gravity Control?

THE GOAL is to mainstream gravity-control technology as quickly as possible. Academia has all the equipment and the expertise, but none of the scientific curiosity, so nothing can be expected from that quarter. This leaves two paths open: angel investors or a ragtag band of engineers in the grand tradition of the Silicon Valley garage. As romantic as the latter alternative sounds, the most desirable approach would be formation of an independent, well-funded aerospace company dedicated to advancing my father's work the final yard across the goal line.

A comfortable level of initial financing for such a company would be in the neighborhood of $20 million. This estimate is based on the initial funding for a 1980s R&D startup of similar size, adjusting for inflation, and adding 25%. I envision a low-profile, industrial park operation that includes...

- Science staff (four full-time for one year, willing to accept partial compensation in stock options)

- Engineering staff (two E.E.s, two programmers, same conditions)

- Administrative staff (two full-time, same conditions)

- CEO – Primary roles: rainmaker, investor relations, Wall Street interface (same conditions)

- COO – Oversees day-to-day operations, smooths over staff conflicts (same conditions)

- Documentation manager (manuals and reports and video record of experiments; web pages)

- Accountant

- Lawyers

- Secure laboratory

- Off-the-shelf equipment for design and testing, including computer modeling

- Custom-made equipment for implementation of Phase I device

- Machined parts design and construction

Phase I of the three-phased operation would concentrate on proof of concept. A small device would be built that replicated the 1994 experiment, but with improvements. This would be achievable in a short time—six months or less. The demonstration would be held for a limited audience of investors.

A convincing demonstration will ensure a much higher level of funding for Phase II, construction of a drone. Uncertainty over engineering issues suggests Phase II would require one to two years. The success of the drone project would lead to Phase III: construction of two manned flying vehicles—one space shuttle, and one private aircraft such as described in *Chapter 5*, p. 19.

The company would become profitable as soon as patentable devices have been constructed, which would probably occur at the end of Phase I. Alternatively, the company could avoid filing, as SpaceX has done, in order to avoid exposure of IP. Based on my father's failed patent filing in 1980 and public disclosure of the gravity control technique in 1981 and in subsequent writings, the raw process itself is prior art and cannot be patented,[55] but as described elsewhere, the specific *applications* of the process, including the secrets of generating the gravity control effect with the greatest efficiency using advanced metallurgy and other techniques, will provide a cornucopia of patents and licensing opportunities. Completion of Phase III will immediately spawn mission-oriented spinoff companies not unlike SpaceX and Virgin Galactic, but with a thousand times the capability at a hundredth the cost.

Then there's the auto industry and fossil fuel companies. Please reread *Chapters 6* and *7* with this in mind. Contract negotiations with these companies should begin as soon as possible, and deals should be closing no later than the conclusion of Phase II. Then there are government contracts, particularly with the DoD, which will be worth billions.

While it would be almost impossible to weaponize gravity control itself, I can see applications in insertion and extraction operations, surveillance, and weapons *delivery*. If I were on the Board of Directors of our hypothetical company, I would vote against any contracts involving destructive use of gravity control. Regrettably, no one company will be able to determine how gravity control is used. However, it is expected that the need for war will decrease as gravity control opens the space frontier and the overview effect seeps into global consciousness. As noted in *Chapter 9 – Geopolitics*, gravity control can be expected to evolve like aerodynamic flight, so there is no reason to withhold it from the market out of misguided fear. On balance, aerodynamic flight has been tremendously beneficial to the world. This will be even more true for gravity control.

Who would want to finance such a speculative enterprise?

First of all, in my opinion such an investment would be a lot less speculative than the three-ring circus I witnessed during the dot-com craze in the late 1990s. The point of this book has been to communicate why gravity control is a good bet, though I am obliged to caution the reader to consider *Chapter 2, DISCLAIMER*, before writing any checks.

A likely candidate to back the company would be one of today's forward-thinking billionaires who

---

[55] See *Appendix F*, p. 269.

finds the message in this book inspiring, someone who is intrigued by the idea of turning the dreams of a real-life Zefram Cochrane—the maverick physicist who invented warp drive in *Star Trek* lore—into reality and going down in history with the inventor as an immortal hero of science and human progress. In my opinion, there is a high probability that this would be the outcome.

The $20 million figure may seem large, and it is, but there are many independently wealthy individuals who invest far more than this in political campaigns that have at best a 50% probability of success, and often worse. Given my father's track record in aerospace and the results of 1994, I regard the probability of success in gravity control to be 90% or better. The critical question is risk vs. reward (see *Chapter 28*, p. 173). This is in part (probably a large part) a psychological barrier based on one word: *antigravity*. In my view, of course, the risk is small, the rewards are astronomical and the word is nothing but a word. My father's lifework was and is *real*.

Investors could pool their money, of course. Information Appliance, Jef Raskin's startup following his departure from Apple, was financed by a small group of investors, including one V.C. from Sand Hill Road. I became familiar with the investor mindset while serving as the employee representative at meetings of the Board of Directors for this company in the 1980s and later in pitching gravity control from 2000 to 2012. Most were risk-averse and would only become interested in gravity control at the end of Phase I, when they wouldn't be needed anyway, but there were a few who might have been interested in backing the technology from conception.

Perhaps the latter comment tips my hand: Naturally I would like to take part in such a company. In all honesty, it could be done without me, but I bring some unique qualifications, evidence of which I hope has been demonstrated in this 282-page résumé. I have known a fair number of CEOs, and I know that I am *not* CEO material. However, I believe myself particularly well-qualified for the role of COO, since I am intimately familiar with just about every other position in the foregoing list, and have a unique perspective on what must be accomplished. I am also familiar with the hothouse environment of a startup, which is somewhat akin to a Roman slave galley, but with Jolt cola in the fridge and a Keurig coffee machine and a microwave oven on the counter. There may be some cachet to having my name on the masthead once Phase I has been successfully completed. My presence would also help with the backup plan for cost recovery if the project should fail (see below).

FLEXIBLE COST OF PHASE I

The $20 million figure above represents an estimate for proper completion of Phase I. Successful completion of Phase I ensures a flood of investor money for Phase II and Phase III. However, Phase I could be accomplished for a lot *less* than $20 million, especially if one of the investors already has a laboratory or access to a laboratory or some of the equipment, especially an EPR device. If some of the participants can be convinced to defer compensation until the end of Phase I, the price tag might be less than $1 million. A figure of roughly $800K is given at the end of *Chapter 19*.

COST RECOVERY

If Phase I is a failure, costs can be recovered by selling off equipment and contracting with Hollywood to do a contemporaneous documentary, something like *The Curse of Oak Island*. The lack of breakthroughs in the search for the Oak Island treasure has done nothing to harm the show, which is now heading into its sixth season (2018), and the chances of success for gravity control are spectacularly better than the Oak Island venture, leading to the chance of future documentaries, making this potentially a lucrative ongoing enterprise for the studio that snags it early.

SUPPLEMENT: F. ALZOFON'S NOTEBOOK

After 2012, I was surprised to find an *eleven*-phase breakdown of a vehicle development program among my father's papers. It was, however, written in November, 1981, so the first eight phases describe the path leading to the 1994 experiments, which had yet to occur at the time of his writing. The title and text that follow are by F. Alzofon.

PROJECT TO BUILD A SPACE VEHICLE USING GRAVITY-CONTROL PROPULSION

The plan is divided into eleven phases:

1) In the first phase, there will be a review of the literature for values of parameters which will be relevant to the experiments, as well as the methods of dynamic nuclear orientation, experimental as well as the theoretical aspects. In addition, in the first phase, the present state of study of virtual processes will be reviewed for whatever aspects are relevant to this study.

2) In the second phase, the experiments and apparatus designs will be formulated.

3) In the third phase, hardware will be specified and lists of apparatus required will be compiled.

4) In the fourth phase, orders will be placed for the procurement of the necessary equipment.

5) In the fifth phase, equipment will be received, certified, and assembled.

6) In the sixth phase, the apparatus will be tested to verify satisfactory operations condition and to generate dynamic nuclear orientation. Tests will probably include a repetition of some experiments with known results to verify that the same results are obtained.

7) In the seventh phase, tests on the samples to be used will be made to determine resonance conditions.

8) The apparatus will be modified to include a device for measuring the change in gravitational force;

the magnitude of this force will be estimated.

9) A working model of a vehicle using the effect found above will be designed and built. It will be tested in a variety of environments.

10) A large-scale vehicle will be designed, built and tested.

11) Interplanetary exploration will begin.

SUMMARY (by F. Alzofon)

The research program proposed will proceed in stages, each of which will prepare the way for the following stage and will itself provide the maximum number of significant results. A structured program will be detailed in which every step will serve as a justification for the following steps; successful completion of each step will be required for the continuation of the following steps.

Accomplishment of the program is expected to require about four full-time personnel and two part-time consultants. The persons acting as consultants will vary with the needs of the program; they will be manned at the level described.

It is expected that the program envisioned [Phase 1 – 8] will require two years to accomplish and will cost approximately $300,000. [According to the U.S. Department of Labor's CPI Inflation Calculator, this would be $876,142 in 2016 dollars. The figure might be considerably lower for an existing laboratory.]

Successful accomplishment of the program outlined will make possible the design of advanced spacecraft which can travel between planets in much less time and more cheaply than is now possible with rocket ships. These craft will be smaller and more navigable than rockets and can remain on trips through space for an indefinite period of time, since fuel can be collected during the trip.

The final report will detail technical accomplishments, describe costs, and will outline the expected uses of the research results.

It is expected that the probability of success of this program [Ed note: all phases] will be close to certainty since the theory deals only with known and accepted concepts of modern physics (although these are put together in a different way than usual). Moreover, only known and successfully used laboratory techniques are proposed to accomplish the goals outlined. Since these techniques will be investigated in areas where they have not been used before, it is inevitable that new scientific data will be generated in the program, even if not all of its stated goals are reached.

On the other hand, since the proposed and expected results are based on analogies which have been used successfully for many years, it is very likely that a large degree of success will be experienced. In such case, the way will be opened for large-scale and inexpensive exploration and colonization of the planets. Raw materials can be mined from outer space and brought back to Earth to replace dwindling resources. It is likely that materials in an already pure state for human use can be found and mined owing to the extreme conditions met in outer space. Manufacturing of products now requiring a vacuum can be inexpensively accomplished in outer space and new products manufactured in zero-gravity will be discovered. In addition, robotic manufacturing processes can be sped up by transferring the zero-gravity effect to robot armatures, since the inertia-free environment will allow running robotic armatures at extremely high speeds without overbanking.

NEXT-LEVEL EXPERIMENTS WITH GRAVITY CONTROL

The following three illustrations depict general concepts for Phase I, Phase II, and Phase III of gravity-control development.

PHASE I - PROOF OF CONCEPT

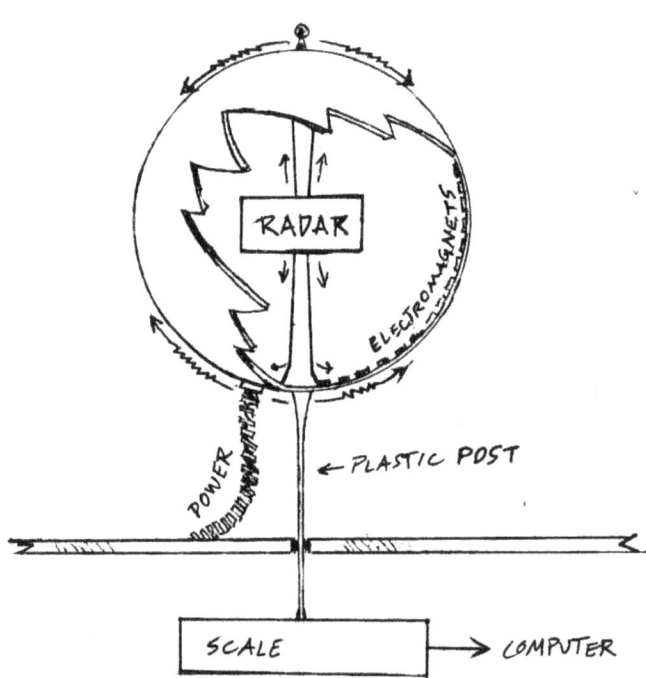

Aluminum sphere sits on post connected to scale. Interior of post contains data line, feeding to computer. Power source is external. "Radar" means "microwave source." Interior of sphere is covered with electromagnets. Minor electronic components not shown.

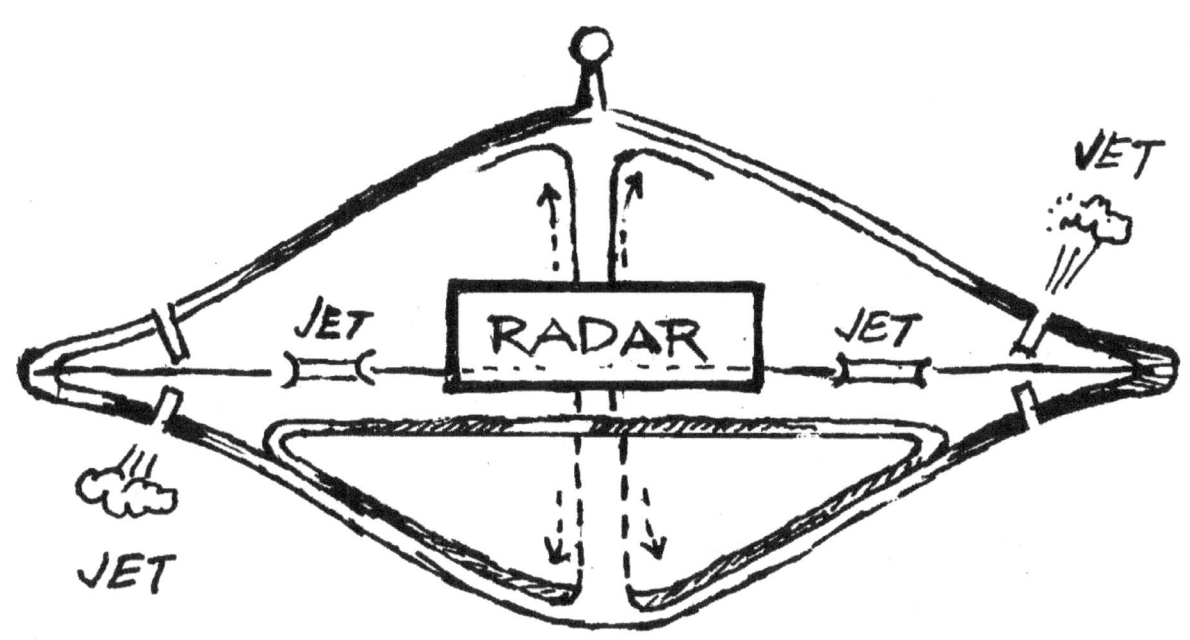

The Phase II drone introduces many new elements requiring considerable R&D. However, Phase II should be self-financing. Remote control is accomplished with wireless signals received by the antenna on top of the saucer. The power source is now internal: a capacitor in the belly of the craft (triangular area below the midline). This will, of course, limit flight time.

Propulsion is accomplished with air jets on the rim. The jets *parallel* to the rim counter the tendency of the saucer to spin. Small jets will create incredible accelerations, but the drone is not intended to be space-capable, since this involves considerable additional engineering.

R&D should minimize microwave radiation on the interior and exterior of the vessel. Experiments should show how to conduct nuclear orientation to the hull via aluminum wires extending from the central microwave source.

Data will be sent back to the ground wirelessly and duplicated onboard the saucer. Much of the maneuvering will be computer-assisted.

PHASE III - MANNED VEHICLE

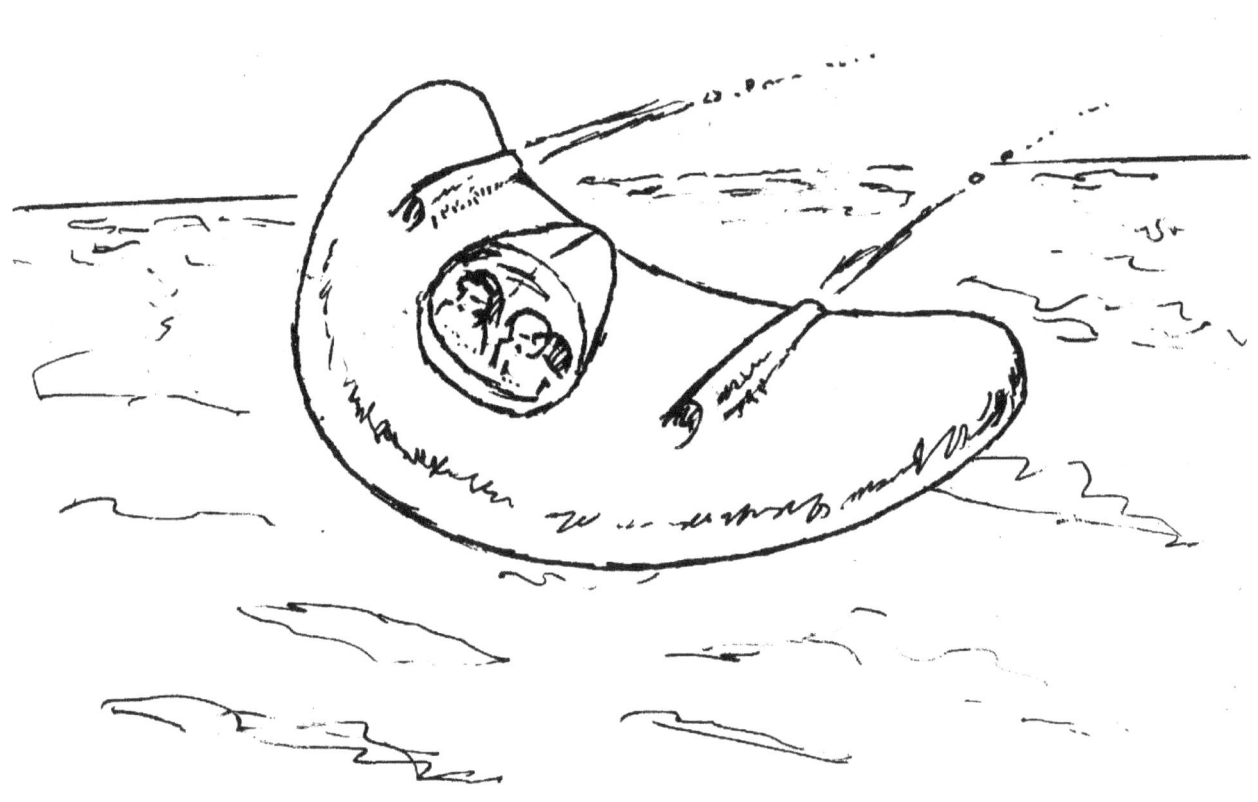

Phase III will sound the starting gun on the new economy. Pictured here is the *Europa*: the two-man, terrestrial air car described in *Chapter 5*, cruising at 50,000 feet. Gravity control is not confined to one hull design: The semicircular, flying wing shape was designed to cater to passenger desire to "face front" when flying.

The size of the turbojets has been exaggerated: small, model-airplane-size jets would be perfectly adequate. A couple gallons of alcohol will get you 1500 miles. The scoops on the front of the jets are "for show" only. They are simply a reminder that the air stream can be diverted to run generators that will keep the capacitors charged. The actual configuration of the scoops would be two thin slots on the underside of the vehicle.

Chapter 27

Real News vs. Fake News

EVENTUALLY, we hope, news will arrive that someone, somewhere, has repeated the 1994 experiment and met with success. The academic community will be skeptical at first, but when the results are duplicated in universities around the world, everyone will acknowledge that the force of gravity has indeed been subdued. While the physicists who conducted the experiments are waiting to collect their Nobel Prize, they will hold a press conference and acknowledge their debt to the UFT. Meanwhile, companies will form to exploit the new technology. The conquest of space will begin in earnest, fossil fuel companies will begin to look into space mining and energy harvesting, Detroit will flock to the new companies seeking designs for air cars and landspeeders. The airline companies will commission passenger planes that take off like a helicopter and fly at twice the speed of sound. Apple, Microsoft, and IBM will scramble to make bids on building software and hardware for the air-traffic-control infrastructure that will be needed. And on and on and on.

This is the rose-tinted, fairytale fantasy. The reality will probably be different. *Much* different. The purpose of this chapter is to prepare the reader for an alternative scenario, such as the one that engulfed "cold fusion" in 1989. In March of that year, University of Utah electrochemist Martin Fleischmann reported production of heat by nuclear processes at room temperature. The press called it "cold fusion." A furor erupted over Fleischmann's methods and results, and by late 1989, cold fusion was discredited in the minds of most scientists. We are not here to argue the case for or against cold fusion. We are merely noting that gravity control is likely to trigger a similar uproar. Whether or not the criticisms are true, it will lead to a clamor in the press, and in the ensuing swirl of accusations, denials, and counteraccusations, the truth may become a victim.

What would motivate such a controversy? Read *Chapter 12, The Air Force Takes an Interest, Chapter 14, Tesla Country, Chapter 30, Theory Matters!* and *Chapter 38, A Word About Expert Opinion,* to understand the environment in which this baby will be born. It is *not* a friendly one. Rather, gravity control will be surrounded with aunts and uncles who would just as soon strangle it in the cradle.

Really? Yeah, really. Gravity—the topic—makes people behave very strangely. When "The Origin of the Gravitational Field" was published in 1960,[56] for example, my dad waited for invitations to speak at universities or to set up a research program. Instead, he got three letters from professors who claimed to have had the idea before he did and had it better. One of them accused him of *stealing*. My father was upset; he had never even met the guy. One of the trio cited a paper that *proved* he had the idea first (he was the only one to do so). Said paper had no resemblance to the gravitation paper.

There was an exception: A group of physicists in Japan wanted to set up a research program. Then they stopped writing. Six weeks later, my dad wrote to them. The head of the university wrote back and apologetically explained that the research group had come down with some kind of deadly virus *en masse*. The entire group had been disbanded and several were disabled. I *said* things got weird.

Things got Gary Larson-ish, too. In the 1980s, a professor was engaged in a conversation with my dad when he got red in the face and hissed, "How dare you even *speak* of such things!" Then he stormed away with nose held high and elbows pumping. My dad said this was a frequent reaction. I didn't believe him until I saw it myself: a Nobel-nominated physicist threw (and I mean "*threw*") "The Unity of Nature" across the table at me in a coffee shop in Palo Alto. As I smacked it down to keep it from landing on the sidewalk, he fumed, "I don't *have* to read this! It's *bullshit*!" Veins knotted on his temples, spittle flew from his lips, his hands trembled, and his face became so red and contorted with venom that I thought he was going to have a stroke on the spot. The episode is reported in depth in *Chapters 24-25* of *The Top-Ten UFO Riddles*. What got him so upset? *Nothing*, it turns out. He was completely wrong. Not that it matters: He asserted his authority, and shut down the discussion.

In the 2000s, I met with many scientists who would not look at the papers. They *knew* there was nothing to it just by looking at the publisher (not a top-tier journal). The fact was that the top-tier journals had printed a similar theory by Hal Puthoff, a Stanford professor and a friend of my father's, but apparently they didn't like it when my dad submitted "The Unity of Nature," even though they'd published him in the past. The differences between the two papers were in my dad's favor, I believe, but no need to elaborate—the *prejudice* was obvious. Engineers, on the other hand, tended to look bored and interrupt me for a lecture about their own pet theories. Only *gravity* inspires this kind of response, by the way. There is something about gravitation that drives mere mortals mad. Can we expect different if gravity control is shown to work?

More often than any of the above, however, the response to my dad's papers was stone silence. The peer reviewers weighed in with opinions that indicated they had never even read the abstract. This is why I can say with some degree of confidence that the readers of this book who've gotten as far as

---

[56] *Advances in the Astronautical Sciences*, Vol. 5, Plenum Press, N.Y. (1960), a publication of The American Astronautical Society.

Chapter 17 know more about the UFT and the technology of gravity control than anyone in academia, even if they can't understand a word of *Chapter 37*.

The point is that academics are *not* a group who will greet news of gravity control with glee. They are not likely to acknowledge it, and if they *do*, it may be to take credit for thinking of it first, or to make it fit the GTR somehow. More likely, however, they will disparage the source, since it will probably be someone outside academia—someone with a pioneering or entrepreneurial spirit who had the temerity to ignore them — vulgar businessmen and engineers, in other words, not *professors*. It's a great disappointment to have to say this, because I love what academia stands for, but every time I've placed my trust in this group, they've behaved strangely and unprofessionally. Based on decades of experience, it would be odd to expect anything different today. Am I bitter or angry, by the way? Nope. Just reporting the facts.

One thing "they" can't gainsay will be experimental success. If things *fly* — hooray! It will be Kitty Hawk all over again.

DISINFORMATION

Certain groups, it is said, have a vested interest in keeping gravity control in a black box. I can't imagine why they'd want to do that when there's trillions of dollars to be made—not to mention the species survival thing? I presume "they" like our species as much as the next person. But let's say for the sake of argument they exist. Such a group might well launch a disinformation campaign — perhaps more than one — to pollute the channels of communication and make the whole thing go away.

If so, expect not just *one* false report, but many. Some will be highly sophisticated and convincing, perhaps endorsed by a physics professor who's famous for something or other, especially in general relativity. Reports of success and failure may appear at the same time and compete with each other for attention, which will sow more seeds of confusion. YouTube, Amazon, Facebook and Twitter will buzz with snarky comments from morons and trolls who aren't qualified to critique a matchbook rocket, let alone the UFT. It is possible, even likely, that these contentious arguments will be manufactured by operatives posing as different people. *Fake news!*

How to differentiate real news from fake? The most reliable reports are likely to come from universities in Scandinavia and other countries not burdened with the problems of empire. Apart from that, the only certain way is to construct the experimental apparatus yourself and throw the switch.

A more sophisticated disinformation tactic would be to claim *success*, but say that it was necessary to modify the apparatus, say, with a massive power source, or all kinds of subtle modifications. Patent applications have already been filed, so why bother? Nothing to see here, move along, and by the way, *fuhgeddaboudit*, too.

A "poison pill strategy" might be employed: A modified setup will be described. It will sound plausible, but if anyone should attempt to duplicate it, it will be a costly dud that will discredit the technology. For example, it may be suggested that pure aluminum works, so no need for colloidal iron inclusions. Or perhaps the only thing that works is an alloy of *platinum and chromium*, or maybe a ruby crystal, since rubies are made of aluminum oxide and their color is due to chromium. But it has to be a really big, really "natural" ruby from the Amazon rain forest, costing $15,000 a carat. Good luck with that. The flood of lies will muddy the water and slow down or discourage other experimenters, and the truth will be buried amidst the conflicting reports.

WHAT IF IT FAILS – FOR REAL?

Reports of failure must be examined closely and verified, since the apparatus has to be precisely constructed and tuned for the test to be valid. If it isn't, the experiment will fail.

Setting that aside, if it turns out that my dad was wrong about gravitation, it would surprise me, since he was right about so many other things and he knew the science better than anyone. It would be painful, since nothing mattered more to him, and he spent a lifetime researching and writing about his theory. But it won't change my conviction that he was a great scientist, and it will not diminish the rest of his accomplishments.

All my life, I've respected his vision of a brighter future created by gravity control: the conquest of space, world transformation, inspiring a new generation of scientists. Failure will cause that dream to evaporate, along with my hope that his invention might have put an end to global warming and shifted human consciousness away from conflict toward cooperation. The future, bleak as it is, will become bleaker, but at least his theory will have been given the fair and impartial test it deserved all along.

MOST DESIRABLE OUTCOME

Without a doubt, the best outcome would be to hear multiple reports of success from experimenters around the world, unimpeded by disinformation, with ensuing R&D projects blossoming everywhere, leading to the technological revolution envisioned in *Book I*.

I would only ask the experimenters to honorably acknowledge the source and contact the author about reviving the 1980 patent filing (see *Appendix F*, p. 269). There will be more than enough fame and money to go around. After all, neither Bill Gates nor Steve Jobs *invented* the personal computer, yet they did very well indeed, as did the stockholders in their companies.

Chapter 28

Risk vs. Reward: A Calculation

THE FOLLOWING is a quick review of risks vs. rewards for the gravity-control project, written with investors in mind. However, we haven't forgotten that the first paragraph of *Chapter 0* contained the suggestion that "*You* might have a hand in deciding the outcome of this saga." Readers who've stuck with us this far are entitled to an answer to the question of "How?" since they are more likely than others to be sensitive to the unique choice offered here, the choice between doomsday and the stars.

Most thinking people these days have the sense that some kind of quickening is in progress. Some rough beast, its hour come round at last, slouches toward Bethlehem to be born. What form that beast will take isn't exactly clear: Climate disaster? Nuclear war? Rogue asteroid? Economic collapse? The menu is extensive. But what can be done? Most of us are powerless. We look after our own affairs, tend our gardens, hope for the best, and ignore the storm clouds on the horizon. We seek the consolation of philosophy or religion, numb our senses with alcohol or drugs, throw ourselves into sports or entertainment, cast ourselves at the feet of some guru or other—all the while knowing that our children will inherit a world that's worse than the one we grew up in. Meanwhile the guardians of the current paradigm fortify their positions, feed us pabulum, and make ready to abandon ship when the deck begins to tilt.

My position is that science and technology got us into this mess, and science and technology can get us out. This is not a hypothetical: Gravity control is real, and it will arrest all these negative trends at once. But the last sixty years or so shows how difficult it is to get this message across. People are stuck on solutions that don't work and feel good, such as those listed above. In the current environment, a single word—*antigravity*—is enough to derail *any* discussion before it even leaves the station. Gravity control needs an angel who sees through the illusions, writes a check, and makes it so. That's all.

Which brings me to you, the reader who doesn't have a few million in spare change. All you have to do is spread the word—mention this book to, say, two other people. You don't have to stick out your neck. Just say "It's interesting," and let people decide for themselves. Write a review on Amazon. Eventually the message will reach the right set of ears and gravity control will have its Kitty Hawk. We are 99% of the way to the goal. All it will take now is one nudge to get us off the ground.

RISKS

- **No endorsement from academia**. More accurately, academia has never refuted or even criticized F. Alzofon's theory. Doing so would require their actually having *read* the papers, which they don't seem to have done. F. Alzofon had all the training and academic credentials one would expect for the inventor of a technology such as this. His record as a problem solver in the aerospace industry was outstanding, and all of his scientific work in other areas was lauded by his peers. Only in the area of gravitation did resistance arise, and if any reason for it was given, it seemed to revolve around the pedigree of the theory, not its content. (See *Ch. 38, A Word About Expert Opinion*, p. 241.)

- **The theory doesn't follow Einstein.** Not true. It is based on Einstein's special theory of relativity, which is one of the most thoroughly validated physical theories in history. In addition, F. Alzofon's unified field theory is covariant and makes many of the same predictions as Einstein's general theory of relativity (see *Chapter 37 – A "New and Simple" Idea*, and *Chapter 30, Theory Matters!*) The 1994 experiments *more* than suggest that the science behind the technology is valid.

- **No demo device.** The 1994 experiments produced results in conformance with the inventor's predictions, as shown by the charts in *Part IV*. Reconstruction of the 1994 experiments is the aim of this book. Stubbornly demanding someone else "prove it out" ignores the evidence and ensures that we will all be sitting on the deck of the Titanic without lifeboats when the ship hits an iceberg, instead of boldly going into space tomorrow. We need to act decisively, *now.*

- **Electromagnetic fields don't affect gravity**. The inventor's unified field theory identifies the physical cause of gravity in subatomic processes that *can* be influenced by electromagnetic fields. The method of interference, dynamic nuclear orientation, has been known since the 1960s and is in use today in the assaying of organic molecules (electron spin resonance). This objection doesn't wash.

- **It's *antigravity*, which everyone knows is science-fiction**. First, it's not *antigravity*—it's *gravity control.* The energy in the gravitational field is not opposed, it is *transformed* and temporarily diverted. It is true, however, that science has been scoffing at "antigravity" for decades. But the scientists who do the scoffing are informed by general relativity. FA's unified field theory is a breakthrough that allows for direct control of gravity. Scientific progress should not be obstructed by word magic.

- **It's a lot of money to bet on a dream**. Frederick Alzofon bet a *lifetime* on this dream, which he took ninety-nine yards toward the goal line. Fortunately, his legacy includes hard data from scientific experiments. The data, plus the strength of the theoretical background and the career achievements of the scientist (see pp. 259 - 267), suggest a high probability of success. During the dot-com craze, the author saw tens of millions bet on half-baked ideas with no track record of success whatsoever. A bet on gravity control is actually *much* more likely to succeed than anything the author witnessed in Silicon

Valley during this era. The amount of capital required to complete Phase I is flexible, as well, and the size of the investment is relative to investor resources. Also, much of the investment can be recovered in the event of failure through sale of assets and possibly a cable documentary (see *Chapter 26*, p. 164).

Overall, the risk is small relative to the rewards when evaluating gravity control. But one must also consider the *external* risks: Without a game-changing technology such as gravity control, we will all remain subject to the current paradigm, which ensures a disaster of Titanic proportions in the near future. The current paradigm is simply not sustainable, and probably not survivable. An investment in gravity control is an investment in a new economy that will keep commerce alive deep into the 25th century by ensuring survival of the species (see *Book I*).

REWARDS

Successful completion of Phase I, the least expensive phase of the project, means the following:

- The enterprise becomes self-financing from then on.

- Company stock becomes among the most valuable on the planet overnight.

- The company will be the first to nail down contracts for the new technology. It will have a huge lead in R&D.

Successful completion of Phase II means

- Contracts with automotive companies, air and aerospace companies, dozens of patents.

- Huge increase in stock value.

Successful completion of Phase III means

- World conquest, in a sense: The company will wipe out competition for space transport contracts and subsidiary industries will be spawned by the dozen.

- Space mining becomes a reality.

- Lunar and planetary colonization gets underway on a massive scale.

- A new Detroit of private transportation vehicles begins to appear.

- Contracts to build the infrastructure for a gravity-control transportation industry will be won.

- Space tourism will become a going concern.

- Global tourism with gravity control will have some of the features of private-jet culture.

- Space colonies in space stations will begin construction.

- Space-based industry will gain a foothold.

- Space-based energy production begins.

- Contracts with the fossil fuel industry to expand their operations into space mining will be won.

- Mars colonies and terraforming Mars become a real possibility

- Subsidiary companies will be founded for construction industry applications, high-speed robotic equipment, cargo loading and unloading, dry dock facilities for oceangoing ships, artificial gravity, etc.

- Opposition to alternative energy will fade as the fossil fuel industry finds a lucrative way out of their current problems (see *Chapter 6*).

- As the overview effect spreads, world culture will change. The emphasis will shift from offensive to defensive weapons and policing space missions and gravity-control air traffic. The military-industrial state will have plenty to do in the new economy, and that means contracts and patents.

CONCLUSION

The inventor predicted that gravity control would be worth the combined GDP of all the industrialized nations on earth within ten years of its debut (see *Chapter 5*). In our opinion, the risks are small, while the rewards are astronomical. The final calculation is yours.

Part V

Theoretically Speaking

Chapter 29

Introduction to *Part V*

READERS **who want to know more** about the theory behind gravity control will find the answers in Chapters 30 – 38 and in the references listed in *Appendix C* (p. 259). The easiest material comes first and the most difficult last. A preview of the contents of each chapter is provided below.

30 *Theory Matters!* – A comparison between F. Alzofon's gravitation theory and others that have staked a claim on the same territory. Includes discussion of the scientific method. (By D. Alzofon)

31 *Gravity and Antigravity* – Transcript of a 1994 conversation. One of the easiest to understand.

32 *Random Thoughts about Engineering* – More of the previous conversation. Nuts and bolts.

33 *Tech Talk with Jim McCampbell* – Discussion of technology with nuclear engineer and author.

34 *A Model for the Origin of the Gravitational Force and Its Control* – Introduction for investors.

35 *Technical Background* – A 1982 introductory essay for potential investors.

36 *The UFT and the STR* – Physicists wondering how the STR could lead to a unified field theory will find an introduction to the answers here.

37 *A "New and Simple" Idea* – F. Alzofon's final paper on the UFT. Postgraduate material.

38 *A Word about Expert Opinion* – "If there was something to this, wouldn't Stanford have done something about it?" This was Steve Jobs' question to the author. The short answer is "No." This chapter is the long answer. (By D. Alzofon)

Chapter 30

Theory Matters!

OUR confidence in the claims of *Gravity Control with Present Technology* begins with our confidence in the scientist behind them, continues through publication of his theory of gravitation in 1960, gathers momentum with his 1981 paper on gravity control, and peaks in the experiments of 1994.

We could continue to enumerate the highlights of Dr. Alzofon's career and the reasons why his solutions to the riddle of gravity and how to control it are credible and unique, but perhaps we should take a moment to discuss the competition. After all, some readers are certain that Albert Einstein solved the riddle of gravitation in 1915, citing the fact that the theory of general relativity (the GTR) has held up rather well ever since. Others believe that Nikola Tesla had a *better* theory of gravitation than Einstein, and would have published it if it hadn't been seized by federal agents and sealed away in a black vault along with his plans for a flying saucer. Some believe that inventor T. Townsend Brown successfully demonstrated antigravity devices in the 1950s, and his technology has been classified above top secret ever since. And there are others. *Many* others. Even a casual glance will show that the field of "antigravity" is flush with competing theories, opinions, and lobbyists armed with dossiers of evidence, including videos of things that float.

In a horse race as crowded as this one, how do we rate the contenders? Often the mere mention of an alternative hero is enough to envelop a listener's mind in a fog of peremptory challenges. This fog is not to be underestimated. Over the years, it has ensured that the number of physicists who've read, let alone taken the trouble to *understand*, my father's papers on unified field theory and gravity control could be counted on one hand. Rather than attempt to refute his theory, almost all physicists have simply walked away from it, often quite literally, with noses in the air and elbows pumping.

Wouldn't it be good to have an objective way to determine the truth? Few would object to the approach my dad suggested in a 1989 grant proposal:

> *At present, no clear model of the physical processes giving rise to the gravitational force is generally accepted. Since the unknown provides an open invitation to speculation, a great number of theories have been advanced to explain the origin of the gravitational force. Many of these theories claim the possibility of eventual control of this force. How, then, can anyone distinguish between promising and unpromising proposals when so many kinds are offered? I believe the criteria of the scientific*

method provide a good and practical guide to how to choose a useful line of investigation.[57]

In spirit, at least, *everyone* is committed to the scientific method. But what does that commitment really mean?

There are several definitions of what science is. For example, mathematics is often claimed to have no relation to the real world, but it is agreed to be a science. I am here, however, limiting the discussion to physical science, for which experiment, as well as theory, are necessary complements.

With this understanding, the scientific method, as Galileo and Newton would have defined it, is comprised of the following steps:

> *9. A correlation between physical events is observed in Nature.*

> *10. A theory is proposed to explain how these correlations occur.*

> *11. The theory is tested by experiment.*

All righty, then! Can we get out of the classroom and into the laboratory now? Well, not quite.

These are hardly enough to describe how physical research is practiced, although the above three steps are often the only three listed. In addition, we must add the following criteria:

> *12. The theory must provide* numerical *predictions; experiment must verify them numerically.*

> *13. The theory must contain the successful theories of the past as special cases.*

> *14. The concepts used by the theory must lend themselves in a convenient manner to experimental design and interpretation. The scientific method is a social enterprise.*

> *15. The theory must be easier to use than competing theories.*

> *16. The theory must be falsifiable, that is, it must be possible to conceive of an observation that would prove the theory, or a hypothesis derived from the theory, false.*

Wow. Is all that really necessary? Well, consider the source: As near as I can tell, it was Professor Victor Lenzen, one of my dad's mentors at Cal.[58] My dad once told me the story of the time he came across a graduate thesis in the library that claimed the universe was a living organism. He repeated the thesis to Professor Lenzen, saying, "I know it's nonsense, but can you tell me why it's not scientific? After all, it's a theory, isn't it?" "Because it's not numerical," said Lenzen, and he listed the five criteria above.

---

[57] For context see p. 205.

[58] The other was Griffith C. Evans (namesake of Evans Hall of Mathematics), who awarded his doctorate.

As anyone can see, the "theory of the universe as a living organism" fails on all five of them.

Who *was* this troublemaker, Lenzen, anyway? Victor Fritz Lenzen was born in San Jose in 1890. He graduated at the head of his class at Cal Berkeley in 1913 and received his doctorate in philosophy from Harvard in 1916, where he studied scientific methodology with Bertrand Russell. From Russell he absorbed the philosophy of logical positivism, which was highly influential among twentieth-century scientists. Lenzen devoted his life to the study of science and was an authority on relativity. Einstein himself hailed Lenzen's essay, "Einstein's Theory of Knowledge," as "convincing and correct" in everything it says.[59] Parenthetically, in "The Philosophical Aspects of the Theory of Relativity," Lenzen said he thought a "better name" for the theory of relativity would be "the theory of the covariance of the equations of mathematical physics." Just file that factoid away for now.

As a scientist, philosopher, and historian, Victor Lenzen understood the streams of thought contributing to the foundation of modern physics as well as or better than anyone alive. These wellsprings are now taken for granted in physics departments all over the world, where professors deploy them in theoretical debate and apply them with great precision to the interpretation of physical research. Multiple generations of graduate students now accept them as scripture. But learning them as he did from Lenzen, my father got closer to the source, where questions about basic assumptions could still be asked. Where academic physicists automatically refer back to Einstein's theory of general relativity, and no further, Lenzen took my father back to its historical roots. Today's physicist would consider this a superfluous and unnecessary exercise, but it was out of these dialogs with Lenzen about the wellsprings of relativity that my dad's unified field theory began to take shape.

As we've suggested, no serious contender in the race for gravity control would reject the scientific method. But *now*, confronted by the eight criteria, they might feel a little squeezed. Meanwhile, it's clear that F. Alzofon's education, career, publications, and ultimately the record of the 1994 experiments look a lot better now. Perhaps this book will convince a few partisans for other candidates to consider a *new* horse and adjust the odds accordingly.

We could begin to grapple with the other contenders one-by-one now, but perhaps it's best to let the reader get a head start. Simply compare the record of your candidate, or *any* candidate, to the eight criteria of the scientific method above and ask how well they acquit themselves.

In many cases, you will find it difficult to begin, because many candidates have no theory of gravitation at all, or if they do, their ideas aren't taken seriously by mainstream science. Hearing this, you might be inclined to side with Robert Heinlein, who said, "Never worry about theory as long as the machinery does what it's supposed to do" (*Waldo & Magic, Inc.*, 1950). Heinlein was an engineer, a breed of cat known for practicality and impatience. For those of the Heinlein persuasion, an interest

---

[59] *Time and the Metaphysics of Relativity*, by W. L. Craig, *Chapter 7*, p. 136

in theory ends where T. T. Brown's or Nikola Tesla's flying machines take off (or *allegedly* take off).

Heinlein could tell a great story, but in this case he was wrong. Dead wrong. *Everything* depends on theory, because in physics, a good theory leads to reliable, numerical predictions about nature, and that in turn guides inventors to breakthrough technologies that perform fresh out of the box. We believe the theory of gravitation described here is *just* such an engineering-friendly theory, while the others fail to satisfy nearly as well.

In short, theory matters! Key to understanding our argument in favor of F. Alzofon's unified field theory and the applied technology derived from it, then, is understanding the word "theory." What *is* a theory, and in particular, what constitutes a *good* theory?

A "theory" is a conceptual framework, a system of ideas, that explains cause-and-effect relationships between phenomena. A *scientific theory*, however, is different from a conspiracy theory or a theory such as "Red sky at night, sailor's delight. Red sky at morning, sailors take warning." Both of these fit the *colloquial* definition of a theory, but lack the rigor of a scientific theory as described in the eight criteria.

Let's consider "electrogravitics," for example. The term, coined by T. Townsend Brown, applies to technology that alters gravity with electromagnetic fields. It is frequently mentioned in connection with UFOs, black-budget projects, controversial Area 51 whistleblower Bob Lazar, and inventions by Brown and others. The technology proposed by F. Alzofon in 1981 *also* relates to the alteration of gravity via electromagnetic fields, but other than that it has *no* connection, historical or otherwise, with "electrogravitics." Rather, it grows straight out of mainstream academic physics.

In the view of mainstream physicists, an immediate problem with electrogravitics is found in the name itself, for it is well-known that gravitational and electromagnetic fields interpenetrate *with no effect on each other whatsoever*. So even if T. T. Brown, R. R. Searl, or any of a number of others *did* invent exotic new propulsion schemes that make things float, in the absence of a solid theory of gravitation to back them up, it is impossible to show just *how* they might be using electromagnetic fields to alter the force of gravity, no matter *what* the results of their experiments might be.

Alternatively, attempts to justify electrogravitic technology with Einstein's GTR — *the* academic standard for gravitational theory — have been tortuous, to say the least. One difficulty is that Einstein failed to unify his description of gravity, inertia, and electromagnetism.[60] Another is that he attributes gravitation to "curved space-time," an abstraction singularly devoid of engineering promise. This is why physicists say gravity control is 500 years in the future. Without a unified field theory that describes the common root of field phenomena in terms of *physical processes*, attempts at electrogravitic engineering, whether led by Einstein or anyone else, will inevitably have dubious outcomes and cover

---

[60] *Encyclopedia Britannica*: britannica.com/science/unified-field-theory

their conceptual cracks and leaks with sci-fi terms such as "wormholes" or "gravity wells."

Incidentally, the appeal of these sci-fi concepts is that they are *visualizable*. The instinct for visualizable concepts is a healthy one. As my father said, "To be of general interest to the scientific community, a theory must not only make numerical correlations between events, but do so in a readily visualizable manner."[61] The problem is that concepts such as wormholes *lack* physical referents. Rather, they hang upon a mathematical model, curved space-time, which is often pictured as a horizontal badminton-net serenely undergirding stars, planets, or black holes, or wrinkled, twisted and towel-whipped in fantasies about alien technology. The badminton-net metaphor is *so* common now, so confidently trundled out by professors, as well as scientific quacks, that it has assumed the status of a real physical entity. More of that later. For now, let us observe that F. Alzofon's theory *also* utilizes visualizable concepts, but they are based on accepted *physical* referents.

"Theory paces technology" is an often-heard maxim in Silicon Valley. It means that possession of a good theory makes all the difference in engineering. Pursuing gravity control *without* a solid, experimentally verifiable theory of gravitation is equivalent to feeling one's way in the dark, but electrogravitics, unfortunately, lacks a good theory. Whether or not there have been conspiracies to suppress electrogravitic technologies, this deficit *alone* would be sufficient to explain why claims of success for this technology have failed to arouse serious interest among academic scientists.

Nevertheless, successes are claimed for electrogravitic devices. This highlights another difference between technology derived from a good theory and technology derived via Heinlein's approach: *Empirical results can be misleading.* Devices employing electrogravitic technology seem to require exotic materials, state-of-the-art engineering, and high-output power sources that produce hundreds of thousands of volts. If so, electrogravitic devices will remain out of reach to all but advanced laboratories. In contrast, the gravity-control technology described in this book is only slightly more complex than a microwave oven. It requires nothing more than wall current and relatively common electronic components, which makes mass-produced gravity-control vehicles a distinct possibility. The impact on society of such a technology would be nothing short of revolutionary. This scenario isn't science fiction: It's the predictable outcome of following the scientific method, step by methodical step, from theory to applied technology.

"But wait! [Fill in inventor here] *produced* antigravity effects in his lab!" partisans will insist. Fine, but where's the proof? The independent verification? The peer-reviewed papers? The publicly repeatable demonstrations? The commercial applications? Where's the bridge between electromagnetism and gravity? Questions such as these usually send claimants running for the shelter of a conspiracy theory. In contrast, we say, "Bring it on!" We'd be *delighted* to rerun the 1994 experiments. It would be cheap!

---

[61] 1981 paper, p. 6; see p. 225, Ref. 2.

Everything some future trillionaire needs is out in the open, and has *been* out in the open — readily available and studiously ignored — since 1981. Absent from our brief are vagaries about the science. As for tales of conspiracy, suppression, and cover-up, as yet there are none. Sadly, institutional inertia alone is more than enough to explain why my dad's work has been ignored for decades.

If a theory exists that supports electrogravitic technology, criteria no. 5 compels us to ask another question: "Does it meet certain minimum requirements for a unified field theory, such as covariance?"

Covariance? What the hell is that, and who cares, anyway? Covariance, which was mentioned in connection with Professor Lenzen on p. 183, is a complex topic, but one needn't be an expert to understand why it is indispensable. Quoting from F. Alzofon:

> *Increased emphasis has been placed on the role of a unification of gravity and other forces since the 1960 Air Force survey,[62] and, indeed, if the objective is to eventually control the force of gravitation, it will have to be done through its dependence on other phenomena and forces. Moreover, in the context of modern physical theory, in order to be convincing, this dependence must be stated in terms of a theory that is relativistically covariant, i.e. containing physical laws that have the same expressions for every acceptable coordinate system: this is required by the special theory of relativity. The latter property is necessary in order that relative motion not introduce effects that are essentially new; for example, an electric charge that generates a purely electrostatic field when at rest relative to an observer, will be observed to also generate a magnetic field when in relative motion with respect to an observer: this feature must be included in any credible theory.*

> *To summarize: A frequent modern theme in the attempt to explain gravitation in terms of familiar concepts is the effort to construct a relativistically covariant theory that describes a unified field: a field that relates the gravitational field to other, well-known fields. Thus, such a theory should include what is known about the electromagnetic field, gravitational field, quantum mechanics, the equivalence of matter and radiation, and can be extended to subatomic phenomena.*

This may make your head spin, so let's put it in other words: Any theory of gravitation susceptible to engineering must be part of a unified field theory, because only a unified field theory will tell us how to exploit connections between gravitation and electromagnetism. But a unified field theory *must* include the special theory of relativity, which means it must be relativistically covariant.

Isn't the scientific method a bitch? Wouldn't it be more fun to play badminton with curved space-time? Unfortunately, we agreed to submit the contending theories to the criteria of the scientific method, and the eight criteria are not a Chinese menu. Ask yourself whether any of the competing theories of gravitation or gravity control meet the minimum standard for a scientific theory, and the

---

[62] See *Appendix B*, p. 225, Ref. 1b; also see pp. 11 – 12

answer, to the best of our knowledge, will be a resounding "*No.*"

By now some readers are beginning to suspect that all this talk of theory is some kind of a trick. For the sake of argument, then, let's grant that inventors such as T. T. Brown et al *have* produced "antigravity" in the lab. There's *still* a vast difference between discovering an interesting effect in a laboratory and trying more or less intuitive ways to exploit it, versus beginning with a theory based on historically proven knowledge and designing experiments around it to test the theoretical model of reality. In other words, by emphasizing *results*, the Heinlein method *inverts* the scientific method. Theory becomes an afterthought, which is why the theories surrounding electrogravitics, while they might appeal to popular audiences, are anything *but* music to the ears of professional physicists.

Incidentally, the theory and technology described here explains why it would be a misnomer to call the technology "antigravity." Gravity control, as described by F. Alzofon, has nothing to do with "negative gravity waves," "gravity wells," or some other mysterious, hitherto unknown force of repulsion that acts against gravity.

So if gravity control, as described by F. Alzofon, isn't "antigravity," what the hell *is* it? My dad will describe it in his own words in the next chapter, but briefly, it works by draining energy from the gravitational field through a technique analogous to cryogenics (the science of generating really, really low temperatures). Emphasize the word *analogous*: Cryogenics works on a *molecular* scale. Gravity control operates on a *subatomic* scale, where we find the causative agents of gravitation.[63] In short, we are *cooling* gravity. The "cooling cycle," which is repeated hundreds of times per second, has a cumulative effect, like the action of a pump, so that the weight of a vehicle can be reduced to almost nothing in a relatively short time period without a large investment of power. This cumulative weight loss was predicted and observed in the 1994 experiments. The whole process is safe, as it does nothing to alter the molecular structure of the hull or the passengers. Switching off the power returns the weight of the vehicle to normal by restoring the normal interaction with the Earth's gravitational field.

Technical details are coming up in *Chapter 5*, so there's no reason to dive deeper than this now. Instead, let's list the dividends reaped from beginning with a useful theory of gravitation:

- *Theory* suggested the analogy with a well-known cryogenic process that led to the invention of gravity control.

- *Theory* predicted the ideal materials for vehicle construction (p. 198).

- *Theory* told us that gravity and inertia result from the same physical processes, so it predicted

---

[63] The cryogenic technique, adiabatic demagnetization of paramagnetic salts, drains heat — that is, removes kinetic energy— from molecules. Gravity control uses an analogous method to drain energy from the gravitational field. Both processes are termed order-disorder transformations.

that inertia would decrease in parallel with weight.

- *Theory* allows us to calculate exactly how much weight will be removed, how fast it will be removed, and how much energy it will take.[64]

- *Theory* predicted the flight characteristics of UAPs, offering indirect proof of their reality and indirect validation of the theory itself.

Without a good theory, inventors and engineers will lack a guiding light. And *only* with a verifiable theory of gravitation in hand can the effects generated by UAPs be explained, let alone duplicated or systematically improved upon with present technology.

The question remains, "What makes a *good* theory?" Part of the answer is in the eight criteria. It took centuries to find the other part, which, in a nutshell, was introduction of operational definitions as part of the conceptual framework of a theory. *Operational definitions* are descriptions of real-world phenomena painstakingly constructed from an inventory of measurable physical elements. This is what allows us to build a cause-and-effect model of reality based on experimentally verifiable evidence. When combined with scientific methodology and historical knowledge, operational definitions form the Sutter's Mill of modern technology.

Scientific progress depended on the merger of two disciplines: The art of framing a scientific theory with operational definitions, and the art of experimental design: for example, isolation of an independent variable and studying its effects on dependent variables. Once this merger took place, humanity's control over nature skyrocketed — much faster than our ability to use our knowledge wisely, as is all too evident from the wanton damage we've inflicted on our environment. But there's no turning back; the genie won't obligingly squeeze into the bottle again. When dangerous technologies hold us in thrall, the best we can hope for is the arrival of a technological wildcard that liberates us from the current paradigm and allows us to sail beyond it into calmer water. Currently, we are so bound up in the paradigm of modern physics that such a wildcard seems impossible. In addition, those who profit by the old order fanatically resist change. But there is one incentive that might entice them to follow the Pied Piper of a new technology: *greater riches on the other side*. Gravity control offers this incentive, a point covered in *Chapters 4* through *8*.

So far we've talked about the importance of theory and the probability that F. Alzofon's gravitation theory is valid, based on the experimental record and its adherence to the scientific method. Now we come to UAPs, which are both a blessing and a curse. The blessing is that the correlations between my dad's technology and the flight characteristics of UAPs are too numerous to be accidental. This suggests that the companion book, *The Top-Ten UFO Riddles*, really does contain the top-ten solutions

---

[64] See *Ch. 18*, p. 119 - 120. In case conservation of energy seems like an issue, see fn. 1, p. 13.

and opens the door to a new technology. The *curse* is the allergy of mainstream scientists to the topic of UAPs, which occasioned the euphemism "UAP" ("Unidentified Aerial Phenomena") which allows them to discuss UFOs without bringing little gray men, alien abductions, government conspiracies, and all manner of high strangeness along for the ride.

The allergy of the academic establishment to UAPs combines with their status as arbiters of truth to make it difficult to discuss UAPs in scientific terms, especially if one begins with a theory that fails to pay homage to Einstein's general theory of relativity, the academic paradigm for gravitation. Since F. Alzofon's unified field theory *sidesteps* the GTR, this is an issue that must be addressed.

The GTR provides a sophisticated mathematical model of gravity and inertia, linking them to a geometric property of space-time. In spite of the fun we've had with the badminton-net metaphor, this *is* an operational definition as far as modern physics is concerned. But it leaves something to be desired: a physical basis, like the rest of physical theory up until Einstein. As my dad put it in an email:

> *The reasoning for Einstein's remark that the metric of space-time is a "real" entity is that the gravitational field is part of it and is observable, and therefore the metric is real. I don't believe in this approach, of course.*

The GTR's use of curved space-time as a causative agent for gravitation is a break with the historic evolution of physical science, but it is accepted today because of its remarkable success at prediction.

A few well-known physicists, such as Andrei Sakharov, Stanford professor and SRI founder Hal Puthoff,[65] and Nikola Tesla have expressed reservations about the GTR in similar terms to my father. Many others grumble for a variety of reasons, but find the top-tier journals slow to publish formal criticism. Some academics are sympathetic to the critics, but not eager to go on record, as generations of professors are now vested in the GTR and tend to think of its problematic features as built into the fabric of reality. Research grants, academic status, publication contracts, and royalties flow from the GTR, and mastering its mathematical intricacies becomes a point of pride and a mark of professional status. *Questioning* its fundamental assumptions simply isn't done anymore in high-level conversation about gravitation. This is why my father never encountered *any* difficulty getting his papers on optics, heat conduction, turbulence, and the Sommerfeld method published in peer-reviewed journals, but ran into heavy resistance with his papers on gravitation. When he was allowed to respond to peer-reviewers, he was quite effective in overcoming their criticism, but opportunities like this were the exception, not the rule. To the casual reader, this muffling of debate by a discipline dedicated to

---

[65] Dr. Puthoff and my father were friends and colleagues at SRI and afterward. Puthoff independently developed a theory of gravitation similar to my father's. See H. E. Puthoff, "Gravity as a Zero-Point Fluctuation Force," *Physical Review* A 39 (1989) 2333. They were mutually supportive, and the similarity was never a bone of contention.

truth may seem odd, if not incredible, but resistance of this kind is routine across many academic disciplines.[66]

Because of the scriptural status of the GTR, its failure to generate applied technology doesn't impress academia. Nikola Tesla, however — who is rumored to have known something about physics — heaped harsh criticism upon the GTR precisely for this reason. I hesitate to speak for my father, as I lack the qualifications, but I think he might have said something like this: *A weak track record in technology is consistent with the absence of a unified field theory that includes operational definitions for gravity, inertia, and other properties of the matter-radiation field.*

The theory offered here ranges from the "paper-napkin lecture" of *Chapter 17*, which I reconstructed from quotes from his other writings through *Chapters 31* through *37*, which include a variety of essays ranging from easy to challenging. Physicists will find *Chapter 37, A "New and Simple" Idea* and the formal papers listed in *Appendix C*, p. 259, most interesting (Refs. 1a, 2, 4, 7, 9). It may be encouraging to know that the UFT solves many longstanding riddles in modern physics, such as the wave/particle problem and the infinities that occur in the Coulomb-type laws, and makes several of the same predictions as the GTR. The point here is that there is enough depth to his theoretical work to satisfy *anyone's* demand for evidence that the present volume does indeed contain "solutions from science."

The objective of this chapter was to convince the reader that *theory matters*. More than that, a good theory is *essential*, because theory guides technological development, and not the other way around. But not all theories are created equal. To play the role of a guiding light, a scientific theory must meet a rigorous set of standards, such as use of operational definitions, inclusion of past theories as special cases, the power to make predictions, and amenability to experimental verification. Also, in order for a gravitation theory to be *useful*, it *must* be part of a unified field theory so that engineers and inventors can exploit the connections between electromagnetism, gravitation, inertia, and other phenomena. A theory of gravitation must also be covariant.

We believe that F. Alzofon's unified field theory meets these requirements better than any theory in existence at this time. Most important, the 1994 experiments were successful: A dramatic weight loss was observed at exactly the frequencies predicted and not otherwise, and the saw-tooth pattern it took was also predicted by the theory. If you understand this chapter, you will understand that this heralded a technological as well as a scientific breakthrough of historic proportions. Though physicists may not be comfortable with discussion of UAPs to be living examples of the technology in action, as data shows that they use the same materials and emit the same electromagnetic signature used in the 1994 experiments.

---

[66] See Thomas Kuhn's influential book, *The Structure of Scientific Revolutions* (1970). Kuhn, who popularized the term "paradigm shift," describes how academia fiercely resists change. Einstein, too, encountered obstructionism.

Dr. Alzofon had a long history of success in other branches of physics and applied mathematics, including optics, heat conduction, and the mathematical modeling of radiation scattering. All of his efforts in these areas were as highly regarded as they were innovative. In aerospace, he showed considerable skill in the scientific method and scored many successes in basic research and problem solving. Gravitation, relativity, and unified field theory were the three areas where he acquired more expertise and invested more time and energy than anywhere else. Given his track record and the results of the experiments, we believe it is highly probable that he accomplished exactly what he set out to do in the early 1940s: implementation of gravity control.

If the historical record holds true, then we are currently living under a set of rock-solid assumptions that time will reduce to quaint museum pieces in the eyes of later generations. While scientific skepticism is wise, just remember that skeptics can be witting or unwitting guardians of our illusions. They were wrong about Columbus, wrong about the place of the Earth in the solar system, wrong about plate tectonics, relativity, wrong about aerodynamic flight, wrong about radio, wrong about television, wrong about flat-screen monitors, wrong about data storage limits, computing speed, and smartphones, wrong about warm-blooded dinosaurs, and wrong about the number of habitable planets in the galaxy, and they never even saw DNA coming. *Wrong, wrong, wrong*, over and over again. With a track record like that, it seems remarkable that anyone takes a skeptic seriously anymore.

But people do. Assuming we are correct about gravity control, the power wielded by skeptics has already had tragic consequences. If the technology had been implemented when it debuted in 1981, we would be living in a vastly different world today. Space travel, asteroid mining (and defense against rogue asteroids), colonization of the moon and Mars, space-based manufacturing and energy harvesting — all would be routine by now. You might even have a GCV in your garage. More important, the glaciers wouldn't be melting, sea levels wouldn't be rising, and hurricanes and tornadoes of unprecedented frequency and magnitude wouldn't be marching across the Caribbean, the Southern states, and the Midwest like avenging angels.

Let's hope it's not too late to wake up and take notice.

Chapter 31

Gravity and Antigravity – A Conversation

ON the evening of Thursday, January 28th, 1993, my father and I returned from dinner at a French restaurant in Corvallis, where a couple of glasses of wine had undoubtedly been consumed. He was in a generous and expansive mood, so I took advantage of the occasion to remind him that he had promised to record a lecture that I could play for potential investors in gravity control. The result has been transcribed below.

By the time of this talk, my father had already given me hundreds, if not thousands of physics lectures throughout my life. He had also taught courses in relativity, advanced physics, and mathematics at the university level, so he was no stranger to public speaking. However, it was *extremely* rare that he ever consented to being recorded. For this reason, some of his greatest physics and mathematics lectures have been lost. Among these was one that he launched into spontaneously on a road trip from Corvallis to Newport that covered the entire history of physics with a clarity I can scarcely describe.

The loss of that unique and brilliant lecture had been weighing heavily on my mind ever since, and in view of how difficult it was to get him to record anything, I was hoping to capture something similar. His responses were well-crafted, but it wasn't everything I had hoped, probably because he was tired. But as I listen, I'm grateful to have any record at all of his exceptional speaking ability.

The following transcript represents roughly ninety minutes of conversation, and though it may sound as if he was reading from a book, there is virtually no editing: the entirety of his remarks were off the cuff. He was also careful to tailor his language to an audience of intelligent laypersons, which is why this talk makes an ideal departure point for the rest of the material in *Book II*, which pursues the same subject matter at increasingly higher levels of sophistication. The 1981 paper, which is not included here, is too advanced for this book, so I made no effort to include it, but it is available online. All publications on gravitation are listed on page 259.

Bear in mind that this talk came roughly twelve years after the 1981 Joint Propulsion Conference paper and a year and a half before the 1994 experiment.

THE DIALOG, JANUARY 28, 1993

DA: How do you explain this mysterious force, gravity? Why has it been such a mystery?

FA: Well, first of all, the nature of gravitation ever since the time of Newton has been known to be pretty clear cut. Gravitational force exists wherever you have masses, and this is speaking on a macroscopic scale, that is, greater than atomic scale. And this is a force that is always attractive and it exists between any two masses. The masses are usually imagined to be electrically neutral, and the force, of course, is always an attractive one.

But in modern times, we have come to the realization that matter is made up mostly of empty space, and that the matter—what we call matter—is limited exclusively to elementary particles. These are particles that are roughly of a diameter of 10^{-13} centimeters. This is about the same size as an atomic nucleus, and of course nuclei are made up of an assemblage of elementary particles, so that you might say now that gravitation, while we do observe it on a macroscopic scale, is characteristic of some property of elementary particles, because mass is limited in its existence only to elementary particles.

So the question, of course, is what property do elementary particles have in common that will give rise to the gravitational force for all of them? Because the thing is, that the gravitational force appears to be the same for any type of mass—for stars interacting with the Earth, or with each other, or cold dust in space, over a wide range of temperatures and types of matter: You have the same force acting, and the same force law. With respect to our modern knowledge, we know there are many kinds of elementary particles and yet there again, there seems to be the same force acting, the gravitational force, no matter what kind of elementary particle there is.

So, my point of view is that the thing that elementary particles have in common, you might say in a way, is simply that they exist. That is to say, that they are stable. And this stability in my theory is associated with the internal structure of these particles, which I imagine to be made up of electromagnetic energy, or *field*, that has somehow become twisted in on itself on this very small scale until a hard knot forms. Another way of saying it is that matter is condensed electromagnetic energy and that electromagnetic energy is dispersed matter.

So this argues a kind of universality of the property of being matter: It's all made out of electromagnetic energy or something similar to it. And if that is so, then there is some reason that the particles which are made up of a random motion of the field that forms them—there's some reason that it holds together and is stable. Now whatever it is, it is unnecessary to know in terms of my theory—simply that it does exist and it is characterized by a diameter which is equal to Planck's constant divided by the mass of the particle times the speed of light: the Compton wavelength. That is a single parameter in my theory which is descriptive of the stability of the particle.

h/mc = Compton wavelength

We imagine the particle to be made of a dense accumulation of energy which is increasingly dense toward the center, less so as you go out toward the outside, until the matter shades off into a field—

the gravitational field. I view the field, the gravitational field, and the matter composing the particle as a single entity. There is a unity of matter and radiation in my theory which is very important for the understanding of the gravitational field, where you imagine that the variation of the energy density is something like a Gaussian distribution, kind of a bell-shaped curve.

Now if you imagine this distribution of energy characterized by that single parameter in the presence of another particle, then these distributions of energy overlap. Now in terms of a single particle's view of its environment, what it sees is that it has gained some mass, and now the condition for stability—which is, as I said, a single parameter, h/mc—that parameter has changed, because the mass is different now: there's more mass-energy, more rest mass-energy, mc-squared [mc^2]. And for that reason the equilibrium condition itself has changed: It's h divided by a different mass now, times the speed of light. And in general, the original mass has now been increased, so that the parameter is decreased.

Of course, the distribution of energy now has to have a smaller diameter, and in order to do that, for equilibrium to take place, the energy has to stream toward the center of the single particle I'm speaking of, or at least there's a tendency for it to do that, and this results in an attractive force on the second particle. The way of codifying and formalizing this tendency is in terms of what is called Le Châtelier's Principle, which is a very general and well-accepted principle, and it says that if there is any physical system in equilibrium and its condition is slightly perturbed, then all the parameters of the system will change in such a way as to restore an equilibrium condition. It may not be the *same* equilibrium as before, but it *will be* equilibrium. In the case we're considering there's only one parameter describing equilibrium, and that is h over mc—and that must change, according to Le Châtelier's Principle, to preserve equilibrium and stability.

This is the origin of the gravitational force.

DA: So now, since we have a foundation for understanding gravity, we're now going to talk about a means of altering the gravitational field.

FA: The alteration of the gravitational field falls directly from the model that I spoke of before. We imagine every particle as being made up of a random variation of energy in its interior. This situation is quite similar to the one in molecular physics, where you can imagine, for example, a container full of a gas. The molecules in the gas are moving randomly and as a consequence they're carrying energy in a random manner from one part of the container to another.

Now if we want to alter the mean energy of this gas, there is a well-known method of doing this. Let's consider a paramagnetic salt, for example. There is a process called *adiabatic demagnetization*, which is used to cool the surroundings of a paramagnetic salt. The process, which is similar to the one that we suggest for altering gravitation, is that a constant magnetic field is applied to the salt. This orients the molecules in the salt in a single direction.

Now after the slight heating that might be generated by applying the magnetic field to these molecules has died away—that is, the heat generated has been transmitted to its surroundings, and all the molecules have come to a common temperature, the magnetic field is taken away very suddenly. That's the origin of the word "adiabatic." These molecules are left standing all oriented in the same direction. However, the surroundings are made up of molecules that have *not* been oriented, and they bombard the *oriented* molecules and disrupt their orientation. In the process of doing that, they do *work* on those molecules, and they lose some energy themselves, and on a macroscopic scale that is observed as a reduction in temperature. This method of cooling is used to arrive at very, very low temperatures.

In the case of the gravitational field, I'm suggesting doing something very similar. What we do is to get a suitable material and orient the nuclei of that material by means of a process called dynamic nuclear orientation. Most of the mass, most of the weight in the object comes from the nuclei, because the nuclei are made up of particles, each of which are about 2,000 times heavier than the electrons surrounding them, so that we expect to get most of the effect by orienting nuclei. The effect I'm speaking of is reduction of the gravitational force acting on the matter that you're using as a specimen.

DA: It seems we've changed from talking about using an electromagnetic field to orient nuclei to talking about *dynamic* nuclear orientation. Can you explain the difference? [*In spite of having read the 1981 paper, I didn't grasp the similarities and differences between adiabatic demagnetization of paramagnetic salts and dynamic nuclear orientation. He is about to clarify that.*]

FA: No, I have *not* changed from talking about an electromagnetic field. The field applied to the molecules is a constant magnetic field.

DA: And that's all that's meant by dynamic nuclear orientation?

FA: That's all that's meant by adiabatic demagnetization as a method of reducing the temperature. Now in dynamic nuclear orientation, what you do is to apply a constant magnetic field to the specimen, and it [*the specimen*] has to be properly chosen of course, and this causes the electrons around the nuclei to process around the direction of the magnetic field. The frequency of precession is called the *Larmor frequency.*

DA: Can you explain what "precession" means, please?

FA: It's very much like the motion of a top that is spinning on the floor and begins to wobble around a vertical axis. That's called "precession." Its axis of spin rotates around the point of contact of the top with the floor, around a vertical axis.

[*Now he continues the description of dynamic nuclear orientation*] We then apply a microwave field of the same frequency as the Larmor frequency to the electrons and this causes them to process so violently that they turn over and they become oriented relative to the direction of the constant magnetic field.

DA: "They become oriented relative to the direction of the constant magnetic field." Are they oriented *in line* with it, then?

FA: It's never quite that simple. Actually it's a distribution of orientations according to the principles of the quantum mechanics. They're oriented in such a way that the projection of the magnetic moment of each electron on the direction of the constant magnetic field varies in half-integer multiples of Planck's constant. However, the distribution of orientations has been shifted in the direction of an orientation—there's more orientation around that direction. As you can imagine, the situation that I'm speaking of is not quite as simple as I've described it. But that's more or less the idea.

The angular momentum change in the specimen due to the flip-over of the electrons has caused the conservation of angular momentum to be violated. To restore it, the nuclei also flip over. Another way of saying it is that the electrons are very close to the nuclei and can interact with them very strongly. Even though the magnetic fields originally applied are not enough to turn the nuclei over, the fact that the electrons have turned over causes the nuclei to turn over. Now the nuclei are also oriented relative to the direction of the constant magnetic field.

Now, in analogy to adiabatic demagnetization of paramagnetic salts, we shut off the microwave field very suddenly. That leaves the nuclei oriented. Their surroundings, however, which are composed of—in the case we imagine, at least—to be the gravitational field, which is due to randomly varying energy, interacts with the nuclei.

The manner of the interaction is not being specified here. I have done a considerable amount of research on it, and the conclusion I've come to is that the gravitational field interacts mostly with the outer fringes of the electrical charge around the nuclei. But it doesn't really matter. If there's any interaction at all, the gravitational field will tend to disorient some of the orientation of the nuclei, and do it probably very quickly. In that case you would expect the gravitational field to be weakened—in effect, you might say its temperature is reduced—and the strength of the magnetic field will also be reduced. So that what we have generated is not antigravity so much as reduction in the gravitational field.

[*Part of the discussion is missing from the recording here. The missing portion must have concerned the slight increase in the gravitational force which is anticipated when the microwave field comes on and drives the electrons into flipping over. The predicted weight increase was observed during the experiment, which constituted a significant corroboration of the theory. The momentary spike in weight is analogous to the momentary heating observed in adiabatic demagnetization discussed above. The transcript resumes in midsentence, with FA discussing what occurs when the field is shut off.*]

...microwave field is shut off. Then is when you expect the gravitational field to be reduced. Any vehicle that depends on a process like that will exhibit a kind of bobbing motion, like a saucer being flipped over the surface of water.

DA: Well, as long as we've shifted to the topic of alien technology and how it relates to your theory—without revealing anything, of course, that is proprietary—why don't you talk about some of the features of UFO flight which your theory explains.

FA: The theory that I have described is semi-quantitative and not very detailed, and yet the great value of the theory is that even without that it is possible to predict effects that can be tested by experiment, and it also can be used to explain a great deal that seems to be known about the alien craft—the UFOs.

In the first place the most promising material to get a nuclei oriented is aluminum, because the aluminum nucleus has a very large magnetic moment. Unfortunately the electrons around the nucleus of the aluminum element shield it from a magnetic field and it's very difficult to use a magnetic field on the exterior to orient the aluminum nuclei.

However, you can orient, for example, the ferromagnetic elements' nuclei very readily. However once you've taken the orienting field away, the orientation of the ferromagnetic material—nuclei—decays very rapidly in something like 10^{-7} to 10^{-9} seconds at room temperature. Now what you can do though, is to put very small particles of iron, or else cobalt perhaps—any element that is ferromagnetic—embed it in the matrix of the aluminum material, and because these nuclei are very close to the electrons of the aluminum nuclei, they can act upon it. You can then orient iron nuclei, those in turn will orient aluminum nuclei through the intermediary of the electron cloud around both of them, and the aluminum nuclei have a very long orientation lifetime. At room temperature about 20 degrees centigrade, the lifetime of orientation of aluminum is about six milliseconds, which is very long. As the temperature of the material decreases, the lifetime of orientation becomes very, very long.

Material from UFOs has been collected—some pieces have been shot away from UFOs, and one UFO exploded and pieces were picked up from that [*the crash occurred in Brazil, in the southeastern coastal city of Ubatuba, in the state of São Paulo, September 14th, 1957*]—and they do indeed seem to have very small iron particles in the presence of aluminum or magnesium. Both aluminum and magnesium have similar properties about long-term orientation lifetimes for their nuclei and high nuclear magnetic moments. However, magnesium has a tendency to explode, to catch fire in the atmosphere—presence of oxygen—and aluminum does not [*debris at Ubatuba was found to consist of 99.99% pure magnesium*].

Aluminum, besides, is a very good structural material and certainly on the Earth, aluminum exists overwhelmingly in the form of a single isotope and is very common throughout the Earth's crust, so it makes a very good structural material in availability and strength and so forth, lightness.

The characteristics of motion of a vehicle due to the pulsing of the microwave field have been mentioned, and there are photographs showing that this does exist. The length of the wavelike motion of the vehicle corresponds to a—roughly that is, roughly—to a time of about six milliseconds, which again is predicted by the theory. In addition to that the nature of the reduction of the gravitational field requires that some other method of propulsion within the atmosphere is needed. That has presumably been seen. Sometimes rocket jets are observed. Sometimes there are rotating discs which appear to have slots in them that can thrust against the atmosphere. These are also observed.

The question of very high speeds has come up, and the very high accelerations, which presumably would crush any inhabitants in the vehicle. But inasmuch as the field that reduces the gravitational force would also affect inhabitants within the structure is concerned, the destructive forces on the inhabitants would not seem to be a problem.

DA: Inertia would be altered along with the gravitational field?

FA: For a long time it has been felt that the gravitational mass and inertial mass must be due to the same mechanism because they are equal. In terms of my theory, inertial mass occurs because there is an internal motion to matter of the sort that I have been postulating. This internal motion results in a rest energy of mc^2, and the internal random motion is however *reduced* in my suggested model for gravitation—that is, gravitational reduction—and if that internal random motion is reduced, necessarily the inertia of matter is also reduced in exactly the same proportion—I *think* in exactly the same proportion. And the origin of inertia due to internal energy is very, very similar to the origin of a kind of inertia in an electric circuit—especially one with an inductive impedance, because this is an analogy to what I'm talking about in terms of rest energy in a particle. In a coil, for example, if a current is flowing through it, a magnetic field is generated surrounding the coil.

Now if the voltage pushing the current through the wire—coil—is shut off suddenly, the magnetic field will collapse inward toward the wire coil and the effect of its cutting the—the lines, magnetic lines of force cutting the coil—will generate an electric current which tends to flow in the opposite direction to the tendency of the original electric current from stopping. It will tend to oppose the electric current stopping. Similarly if you begin to generate an electric current in a coil, the magnetic field will expand outward and a counter-voltage will be created which causes a current to oppose the original impressed current. So that no matter which way the current tries to go, there is an opposing inertia—a sort of inertia—that prevents it from flowing in the direction it wants to go, and this is an internal motion which acts just like an inertia.

More to the point, perhaps, is another model in which you have molecules placed in a container, and the molecules have an internal motion and they bounce off of the walls of the container. Now if you try to move the container rather rapidly in one direction or another, the molecules will strike harder on the surface moving toward them and less hard on the surface moving away from them, of this container, so that they will tend to hinder the motion of the container.

Again, that is due to the internal motion of some object within the container, or within the system, and this creates a kind of inertia. Similarly the random motion within elementary particles is considered to resist acceleration imposed upon them in the same way. But this, as the model of the gravitational force has illustrated, this is the *same* process as causes the gravitational force. That is to say, the random motion within the physical system is responsible for the gravitational force and for inertia, and in both cases the effect is proportional to what we call the mass or the rest energy.

DA: So this concludes our question and answer session for today, Thursday, January 28th, 1993.

I could see that my dad was getting tired, and not wishing to put him under any pressure, I drew the conversation to a close.

Chapter 32

Random Thoughts About Engineering

THE SECOND question-and-answer session transpired on the following day. My dad's enthusiasm for talking had considerably lessened, as can be seen by the brevity of his responses and the duration of the conversation itself. Nevertheless, it produced some answers that won't be found anywhere else.

RECORDED JANUARY 29th, 1993

DA: The question for today is to envision a program to build a spaceship and imagine some of the project goals and pitfalls that would inevitably occur. Oh—*money is no object*.

FA: You'll have to ask some specific question.

DA: Okay, what would step one be? Design the hull?

FA: Oh, we'd have to enlarge the scale of the model we built. Presumably we've built a model that works, but now we have to increase the size of the ship. And the phenomena that we rely on may not be increased in the same proportion as the size of the ship, so I guess we have to start the design and do some theoretical work to see what magnitudes we can expect for it. What the power has to be and how much of an effect we can expect. Is that what you wanted to know?

DA: Sort of. What are the main components of the ship? At least give me that.

FA: I can't tell you that, David.

DA: You've already put it into print in a paper.

FA: Well, let's say antigravity screens, how's that?

DA: What are those?

FA: Aluminum screens that you bolt into place on the surface of the vehicle, and the power plant … [*inaudible*] successively orient the nuclei and then orient them again.

DA: Is there a way of testing the orientation in real time?

FA: I don't know.

DA: Okay, so we need a power plant.

FA: Uh-huh.

DA: We need shields...

FA: Screens.

DA: Screens, thank you. Why do you call them "screens"?

FA: Because they're sheets of metal, and screens look like that.

DA: And you talked about generating the effect in a central core and running it out on wires to the screens?

FA: That was one possibility I considered, yes.

DA: It's as good as any. We need a cabin.... How much protection do the occupants need from the microwaves?

FA: Well, we'll include shielding from the microwave field.

DA: Is that a shell that surrounds the cabin?

FA: Yeah, mm-hm.

DA: Okay. Now how come that part of the ship gets involved in the effect, and it's not *isolated* from the effect [*by the shielding*]?

FA: Oh, it *is* isolated from the effect.

DA: Doesn't that mean then that the accelerations will toss the occupants around?

FA: Oh, I see what you mean. I thought you meant isolated from the microwaves. No, it's *not* isolated from the effect [*gravity/inertia reduction*]. The field generated will tend to cancel out, reduce the mass of the occupants. That's true. Uh-huh.

DA: Okay. Do we need special uniforms or suits for the occupants?

FA: It'll help if they [*the flight suits*] have a fabric with aluminum threads woven into them, I think.

DA: Would that conduct the effect to the molecules of their body, or would that simply help shield them from microwave radiation?

FA: I think it'll just shield them from microwave radiation.

DA: Okay. Do you think that designing the hull will be a difficult problem—I mean from the standpoint of strength, temperature resistance, uh...?

FA: Well, we'll build it like a submarine hull is built. How would that be?

DA: The submarine hull withstands pressure, but it doesn't have temperature problems?

FA: Well, a submarine hull is built for strength—structural strength. That's about all I know at the moment. I can't predict what kind of stresses it'll be subject to.

DA: You said something about thermal activity lessening the antigravity effect.

FA: Hm, yeah. The constant switching back and forth of orientation and disorientation probably will generate heat—how much, at the moment I don't know, and how much dissipation of heat will be necessary, I don't know either.

DA: Passing through the atmosphere would also generate heat.

FA: The indication—and it's not very clear—the indication is that the ionized layer of gas surrounding the vehicle will minimize the friction due to the atmosphere on the metal hull. I can't say very much about that, either.

DA: Okay.

FA: The other thing you can do is just go up out of the atmosphere before you turn on the high velocity.

DA: Go slow?

FA: "Go slow?"

DA: Up out of the atmosphere slow?

FA: Yeah.

DA: Okay. Um, now you need some kind of propulsion.

FA: Well, rocket jets would do very well. You can also install a rotating rim that has slots in it, and the slots have angled edges so that you can push against the atmosphere.

DA: What, kind of like a *windmill* effect, or...?

FA: Something like that, yes.

DA: How does that steer you, though? I can see rockets steering the ship, but I can't see the...

FA: You tilt the vehicle in the direction you want to go, and that would steer you. Uh, how do you tilt the vehicle? You perhaps increase the weight on one side over the weight on the other side.

DA: Okay. How long do you think it would take, with an unlimited budget, to build this thing?

FA: Well, the problem would be to get skilled workers. I'd say maybe five years.

DA: Well, that won't do for a novel. Have to do it in one year.

FA: That is not very practical.

DA: I realize that—we're going to *lie!*

FEA [*unamused*]: Okay. You can perhaps get a water tank and install the equipment in that—or a gas tank.

DA: What kind of a.... Well, that's not what I'm thinking of. What kind of a steering mechanism do you think would be optimal, for the pilot? A joystick? A...

FA: Oh, you mean what it would look like in his hands?

DA: Yeah, in his hands—the control mechanism?

FA: I guess a joystick would be a good idea, yeah.

DA: What about up and down motion? A joystick is four-directional.

FA: Mm-hm.

DA: What about movement up.... Maybe you could lift it and pull. Maybe it could be lifted and pushed, and, well, anyway....

FA: I can't think of anything.

The tape runs out here, with many questions left unasked and still more unanswered.

Chapter 33

Tech Talk with Jim McCampbell

INTRODUCTION

The following is a transcript from the tape of a rare meeting between my father and Jim McCampbell at McCampbell's home in the Belmont Hills, near Crystal Springs Reservoir. I wasn't present at the meeting, but my dad made the tape at my request because I believed that with McCampbell's pull at the Department of Energy and other organizations, an experiment was bound to be in the offing, and I wanted to document the events that led up to it for history if it was successful. As it happened, none of McCampbell's efforts bore any fruit and the experiment came about through entirely unexpected channels six years later.

THE DIALOG, MAY 19th 1988

FA: We're at the home of Jim McCampbell. We're discussing the theory of gravitation and how it might apply to observations, explanation of UFO behavior. We have already been talking about the theoretical foundation of the gravitation theory, and I'll have to supply that later. That is the part that will be missing from this tape.

JM: And what we want to do first is to work through an agenda, and I'll mention the items I have noted.

FA: Okay, these are JM's items.

JM: Right. I want to call your attention to the British position in all of the questions afoot—which I'm sure you've had no chance to, you know...

FA: No, I don't know anything about it.

JM: Okay. Well, I can hit that easily and quickly. Then, by the reduction of mass of the machine, based upon your theory...

FA: Mm-hm.

JM: ...then it's quite easy to cause it to move the way you want to. In other words, the forces requirement is surprisingly small. And I've made some calculations along that line.

FA: Mm-hm.

JM: And they're revealing, and in that context there is a tremendously important paper called "Electromagnetic Propulsion without Ionization." It came out of the *Journal of Spacecraft and Rockets* in 1981, and I noted it at that time. I threw it in here for—couldn't do anything with it, but now it makes every bit, all the sense you can imagine.

FA: Hm, interesting.

JM: And then.... Okay, then I want to review the Gulf Breeze case in Florida.

FA: [*perks up*] Yeah.

JM: You take the *MUFON Journal*?

FA: Oh sure, and that's fascinating. I was hoping we could talk about that.

JM: Okay. I have a videotape prepared by a local television station in Gulf Breeze, and it's quite interesting. So I'd like for you to look at that. And then I'd like to discuss the physics of the light source underneath the UFO. Then there's the question of the MJ-12 documents, uh, the Spielberg-Reagan connection...

FA: Spielberg-Reagan?

JM: Yeah. Spielberg is the movie producer.

FA: Yeah, I know. I didn't know there was any connection between them.

JM: Okay. Then I'd like to discuss John Lear and the Groom Lake...

FA: Groom Lear and what?

JM: And the *Groom Lake*.

FA: Groom Lake?

JM: Yes. Groom Lake is the region north of the atomic test site down in Las Vegas. I'd like to discuss...

FA: Who is John Lear?

JM: He's the son of Bill Lear.

FA: The industrialist?

JM: Yes.

FA: Huh. Well, he, well, he's...

JM: [*interrupts*]Then I'd like to discuss Lockheed Corporation and my recent correspondence with the director of the research lab, and the Burbank operations and Colonel Edwards, of, uh, Albuquerque.

FA: Well, in comparison, my proposed agenda looks weak and puny. What I'm primarily interested in is trying to relate, um, the parameters of observations on UFOs to the theory that I'm proposing, and, uh, see if we can gain an insight into how the gravitational field is altered.

JM: This is, this is *precisely* the focal point of my interest.

FA: Well...

JM: It's *exactly* where, where I want to get!

FA: Then this is a very happy circumstance.

JM: And hopefully by the end of our session, now or some extended next week [*sic*], we might get to the point where I can feel I understand exactly how one should proceed to actually manufacture some vehicles.

FA: Well, there are two aspects of my agenda. One of them is what I already told you. And the other one is somehow to get the resources to do experiments so that we *can* build these vehicles so that we can make our own shuttle vehicles. I think of the UFOs as shuttle vehicles more than anything else.

JM: Is that your opinion?

FA: Yeah. You want to use it?

JM: No. Until I find another one.

FA: Basically, moving around in deep space between aggregations of gravita-, of *mass* is not difficult. If you have enough power, you can just—say you have a nuclear power [*plant*], you can gather matter anywhere in space, and there's always matter around somewhere, and you heat it enough and push it out in the direction opposite to where you want to go, and you can build up tremendous velocities. But there *is* a problem in getting off of a planet or other gravitational field—uh, *body with* a gravitational field—once you get close to it, and that to my mind is what the basic uses of the UFOs are. That is why you have the mothership and the small ones. The small ones are the shuttle craft. And

207

the big one, I think, is just to travel deep space. It's big—it doesn't have to be small—a small push will move any amount of mass out in deep space.

JM: All right, let's add to the agenda the present work on your experiment so that I can understand what it is you are trying to do with that experiment.

FA: The experiment.... It's very simple. What we have to do is to get some magnets—I think I spoke to you about it on the phone—we have to get magnets of the sort that they have in nuclear magnetic resonance devices. [*gap*] Get a specimen of very pure aluminum...

JM: Which you now have, right?

FA: I have. And it may or may not be doped with chromium or iron...

JM: ...particles. What size particles?

FA: Micron.

JM: Would the normal process involve taking the sample to some machine shop or foundering shop?

FA: I don't think so—just a small piece of it.

JM: Well, what... to have it melted and have the iron inclusions blended in and then rolled...

FA: You can...

JM: ...rolled out in a sheet?

FA: I don't.... Have you ever seen very, very pure aluminum? It's fantastic. It's grayish blue. It's bluish gray—exactly the color that has been reported so often—and it's very soft. All you have to do really is get some of this colloidal iron powder. You can get it from rare metal manufacturers. And, uh, sprinkle it on the aluminum, and it will adhere. If it doesn't adhere enough, you can put it in a vice and squeeze it tight to force the iron into the aluminum—'cause, you know, iron is hard, the

aluminum is soft—and that's all that would be necessary. Preparation of the sample is the least of my problems.

JM: What's the form of the sample now? Is it a block? Or...?

FA: No, it's a sheet. About three inches by three inches.

JM: Is it quite thin?

FA: I'd say it's about a sixteenth of an inch thick.

JM: And is that thickness suitable for your test?

FA: Sure. And we put that between the pole pieces of the nuclear magnetic resonance magnets. Now the reason for using those magnets is because the magnetic field is very constant. If the magnetic field varies, then the sample will move through the field. If the iron particles get magnetized, you see, and they *will*, then the whole thing will move between the pole pieces, and if the field is very constant, they will *not* move. Okay? We have to separate the effect we want out from that. So that gets rid of a lot of interference. Then we irradiate the sample with three gigahertz microwave radiation—pulsed. The radiation pulses—I put the characteristics down in the paper that I had...

JM: Yes. Yeah, I thought we ought to compare those numbers in detail with the deductions which I made and published in the last paper.

FA: That is the most useful thing we can do. I picked those numbers because they are physically realizable, for one thing, in terms of present technology, which is a little surprising. And the observations that you reported in your book on ufology.

JM: Okay, I'm talking now the data that I summarized in that paper on the effects on people.

FA: Yes.

JM: Okay, so that is a rather broad overview that leads to more precise knowledge of the individual parameter values.

FA: Yeah. You gave in that paper the magnetic moment of the UFO. Of course what we're interested in here is not...

JM: I'm talking about the pulse width and the pulse rate.

FA: Oh yes, you did that, too. But we'll have to know the constant magnetic field too.

JM: Okay, we can get some values on that too.

FA: Yeah. That I would like to see.

JM: And I wanted to know what the magnetic field is in these medical, I mean the nuclear magnetic resonance equipment—what field strength is used.

FA: Why, I gave it in here. It is attainable in the nuclear magnetic resonance [*inaudible*]. I think it's somewhere in the neighborhood of a few thousand gauss. If you're talking about empty space between the pole pieces, then that would be *oersteds*, not gauss. *Gauss* is the field inside...

JM: Inside of the object.

FA: In the solid, yeah.

JM: Okay, that's...

FA: Present technology is, *with* present technology—that is what I wanted to emphasize.

JM: Okay. Find in here the reference to the parameters of the electromagnetic field, and I'll go get a copy of my paper—we'll talk about those points. [*pause*] Here are two copies for your sons.

FA: You sent me one.

JM: Yeah, those are for your sons.

FA: Yes, I understand. We'll be seeing them in the next couple of days. [*pause*]

JM: Okay we're talking about exactly the same thing. We have a strong magnetic field.

FA: Here you are.

JM: All right.

FA: 10^6 oersteds. Oh, no, no, I'm sorry. That's "the nucleus experiences a magnetic field." No I don't see it here, but as I recall...

JM: I'll have a look. I read this yesterday [*pages rustling*].

FA: All I said here is "less than 10^4 oersteds." And let's say something like 4,000 oersteds. And this was at temperatures of less than a tenth of a degree Kelvin. Now of course the experiments that have been done with dynamic nuclear orientation have been done at very low temperatures—a tenth of a degree Kelvin. And the reason is that they wanted to study the relaxation times of the electrons, for example—or the protons, or the nuclei—under conditions where the thermal oscillations, the thermal perturbations, were frozen out. Okay? But otherwise the lifetimes with respect to thermal fluctuations of the nuclear orientation is of the order of...

JM: [*inaudible*] As I recall, having read it. I can't quote what you said about it, though.

FA: Well, it's extremely small. It's microseconds or nanoseconds. Oh, here we are: 10^{-10} seconds.

JM: Your microwave input has to reconstruct...

FA: What is?

JM: The microwave input has to reconstruct energy that's lost—is that you what you said?

FA: No, the point is that the rare earths—of which aluminum is one of the group—are unique in that their lifetime against thermal fluctuations is very long, even at room temperature. It's of the order of 6 *milliseconds*. And that's enormously long compared to 10^{-7} [*sic*].

JM: Yeah it is enormously long. I understand.

FA: That's what makes aluminum unique.

JM: There is the data. I put those marks there yesterday.

FA: Oh yes, here it is. Uh-huh. Fixed magnetic field is 660 oersteds. Pulsed oscillating field at 3,000 megahertz.

JM: Okay, now the pulsed field lasts two microseconds. So that two microseconds is the duration of your pulse. Okay, now let's talk about that. What I found out was that the hearing response had to be between 200 and 3,000 megahertz. That's the cutoff. You're just not going to hear anything outside that bandwidth. So we're okay on the 3,000.

FA: I see.

JM: All right. Now, but I've also found out that the response is, on log paper looks like this.... The human hearing response as a function of the frequency. Here's the frequency. There's a very steep mesa, like that. [*inaudible*] Hertz down here at one....

FA: That's okay.

JM: ...sixty, and a hundred. So that when people are near the UFOs and "hear" the electromagnetic radiation, it's most likely that the pulse width lies between this ten and forty. [*inaudible*] It's most likely, but it wouldn't be possible to hear if it was at the threshold of the hearing capability. So I think the indication is that it's more likely that this is the range.

[Ed note: A long digression on how to interpret JM's hand-drawn chart, which is unavailable, begins here and the tape eventually runs out.]

Chapter 34

A Model for the
Origin of the Gravitational Force and Its Control

INTRODUCTION (by D. Alzofon)

In 2000, I returned to the Bay Area to work with Silicon Valley guru Jef Raskin on a new kind of Internet browser interface that Jef thought would take the world by storm. As is typical in the Valley, the workweeks varied between fifty and sixty hours. What little spare time I had was spent trying to find venture capital to invest in development of gravity control. It soon became evident that I needed a brief but convincing explanation of the technology for an intelligent layperson, so I asked my dad to write an essay that I could use both as a set of talking points and a handout. The result will be found below. Much of the language is repeated here and there throughout *Book II*, but its conciseness and simplicity make it a convenient stepping stone between the preceding chapters and the more complex material ahead. Everything that follows was written by F. Alzofon.

INTRODUCTION (by F. Alzofon)

Although the universal gravitational force law was published by Sir Isaac Newton in 1687, the physical process giving rise to the force has remained a topic of speculation down to the present time. The law itself was simple in form: It depended on the masses (quantity of matter) of the interacting bodies and their distance of separation, and a universal constant independent of the nature of the interacting bodies. Contrary to the manner in which most material interactions take place—i.e. by contact between the interacting bodies—the gravitational force is exerted at a distance and does not require contact between the interacting bodies. It is not localized in space.

To be sure, there are other nonlocalized (or field) forces: electrical and magnetic force fields are cases in point. Like the gravitational force, these are nonlocalized and do not require contact between interacting material bodies. Bodies interacting via electrical and magnetic force fields can be shielded from one another (i.e. neutralized) by use of metal screens. In contrast, the gravitational field can *not* be neutralized by use of metal screens.

In the following discussion, we shall present a model for the origin of the gravitational field and a means of controlling the field (i.e. increasing or decreasing the force) using only present technology.

MODEL FOR THE ORIGIN OF THE GRAVITATION FIELD

As mentioned above, the gravitational field occurs in the presence of matter and not otherwise. Moreover, the law of force (and therefore its origin) is the same for an extremely wide range of physical conditions. For example, Newton's law is the same for the force between the Earth and the sun, as well as between the Earth and the Moon. It is also the same for the cold dust in outer space and between very hot or cold stars. Thus it seems reasonable to seek the origin of the force in some feature that matter in all its forms shares in common.

Modern physics tells us that matter is composed of elementary particles with diameters on the order of 10^{-13} cm. These particles exist in a bewildering variety—electrons, protons, neutrons, mesons, quarks, etc.—and since matter is confined to these and does not occur in any other way, these particles must in some way be responsible for the gravitational force. The population of these particles is different for stars and colder bodies like the Earth, yet the gravitational force law holds true irrespective of particle populations, which suggests that we should seek the origin of the gravitational field in some property common to *all* elementary particles, irrespective of type. The model proposed in this essay is that the gravitational force is a result of the *stability* and *internal motion* of all elementary particles.[67]

To see how these two properties are related to one another, let's examine Albert Einstein's famous relation between matter of mass m (say, of an elementary particle) and the equivalent amount of radiation energy mc^2 (c is equal to the speed of light).

This relation is a way of saying that radiation and matter are equivalent: that is, radiation is dispersed matter, and matter is condensed radiation. On a subatomic scale, this assertion has been proved experimentally: one can be transformed into the other. In turn, this says that every elementary particle has an internal energy of motion of some kind, and this is released when it is changed into radiation. Evidently this internal energy is restrained within a very small region, estimated by the Compton wavelength h/mc, where h is a very small quantity, the *Planck constant*.

There must be two opposing forces acting on every element of an elementary particle: one is the internal motion which tends to blow it apart into a nonlocalized radiation field, and the other a force which holds it together and makes it stable. From the two relations given, something else can be said about the nature of the forces. If m increases, so does the amount of internal radiation-equivalent energy of the particle. But, at the same time, the Compton diameter *decreases* in order to hold the

---

[67] Ed note: So much for gravitons.

214

particle together: the restraining forces become larger, too.

The Compton diameter is only an approximation of the extent of an elementary particle. In actuality, the energy of the particle extends beyond this limit. The figure below is a schematic representation of the energy density variation and not to be taken too literally. The height of the curve represents energy density, while the horizontal line at base represents distance. In other words, the energy density increases at the center of the particle and decreases farther away.

$$| \text{<--------------} \; h/mc \; \text{------------------>} \;|$$

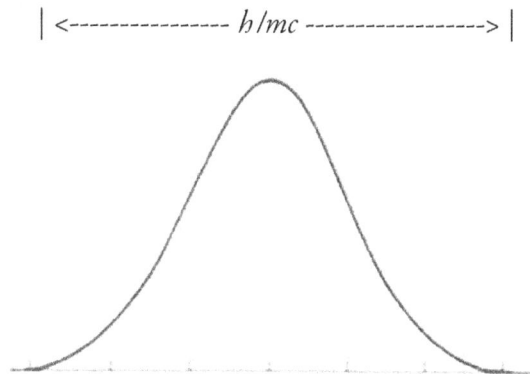

The extended field surrounding the particle is asserted to be responsible for the gravitational force. To see how this occurs, consider what happens when two elementary particles are neighbors, so that their extended fields overlap.

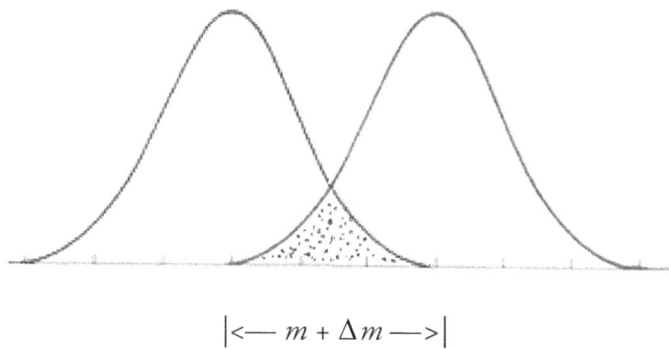

$$|\text{<---} \; m + \Delta m \; \text{--->}|$$

Insofar as one particle is concerned—say, the one on the left—it has gained in energy by a small amount, from mc^2 to $(m + \Delta m)c^2$, and the condition for stability has altered from h/mc to $h/(m + \Delta m)c$, or approximately $h/mc - (\Delta m/m)(h/mc)$. That is, there is a tendency to draw the added energy closer to the central portion of the particle. The second particle (on the right) senses this tendency as a force upon it as a whole. When multiplied by the enormous number of particles making up macroscopic matter, this minuscule effect becomes, in aggregate, what we call *the gravitational force*.

ALTERATION OF THE GRAVITATIONAL FORCE

The model proposed above can be used to design a means of reducing—or increasing—the gravitational force. A clue as to how this might be done is that the internal motion of energy within the particle does not cause a wholesale movement of the particle: every motion in one direction is matched by one in the opposite direction. This is characteristic of chaotic motion with zero net-displacement. This chaotic motion can be altered to a systematic swirling motion for some elementary particles, causing an observable magnetic moment, by placing the particle in a magnetic field. That is, in some particles the internal motion that is responsible for the gravitational field can be given a directional property.

Based on an analogy with the adiabatic demagnetization of paramagnetic salts—a process used to lower the temperatures of these salts to a fraction of one degree Kelvin—we can imagine a procedure for lowering the energy density of the gravitational field (and therefore reducing its strength) by making use of *magnetic moments*. First, the process for lowering the temperature of paramagnetic salts, in schematic outline:

Electromagnetic Cryogenic Process (Adiabatic Demagnetization of Paramagnetic Salts)

1. The paramagnetic salt is placed in a constant magnetic field. This orients the molecules of the salt along the direction of the magnetic field due to the magnetic moments of the molecules.

2. The salt specimen then comes to the same temperature as its surroundings.

3. The magnetic field is quickly removed, leaving the molecules of the salt momentarily oriented.

4. The material surrounding the salt, whose molecules are still moving chaotically, interacts with the salt through these molecules, causing the salt molecules to lose their orientation and also move chaotically. In so doing, the surrounding molecules lose energy and become cooler, i.e., their temperature drops. The lower temperature is communicated to the salt specimen.

5. Steps 3 and 4 are cycled over and over, "pumping" heat out of the salts and achieving extremely low temperatures.

Analogously, it is proposed to use a process called *dynamic nuclear orientation* to orient the electrons and nuclei of a suitably chosen material making up the skin of a vehicle. When the source of nuclear orientation is suddenly removed, the gravitational field surrounding the vehicle will restore the oriented elementary particles to a chaotic state and lose energy in the process. This will lower the energy density of the gravitational field and weaken it. Applied cyclically, this process can be used in vehicle propulsion.

Dynamic nuclear orientation is a known technology, but little recognized outside of the physics laboratory. Its use in gravity alteration requires additional equipment and precise application, which explains why gravitational effects have not been observed in connection with it before this time.

EXPERIMENTAL RESULTS

Preliminary experimental tests of this process have been successful in showing that increases and decreases in the weight of a suitable specimen can be generated by a cyclical process involving dynamic nuclear orientation. Results accord with the numerical predictions of the theory. Weight reduction in the vicinity of 80% was observed.

Chapter 35

Technical Background

INTRODUCTION (by D. Alzofon)

The following document, dated August 14th, 1982, was written in response to Jim McCampbell's request for a general description of gravity-control technology that could be part of a presentation for potential backers. Eventually McCampbell contacted the Department of Energy (pp. 84 – 85), and my father submitted a formal research proposal to the agency. The DoE proposal has been lost, but the following probably represents a draft of at least part of it.

THE NATURE OF GRAVITATION (by F. Alzofon)

Newton's gravitational force law has been verified in a variety of laboratory-scale experiments, as well as in its accurate prediction of the motion of heavenly bodies (including interplanetary probes). In general, it is found that, although the gravitational force varies with distance in the same way as electrostatic and magnetostatic forces, it is measured in the laboratory to be much weaker than electrostatic or magnetostatic forces. For example, the electrostatic force induced by rubbing synthetic fabric together (e.g. in an automatic dryer) is easily observed, but the force between two larger masses, owing to their gravitational attraction—for example, between an ocean liner and a passenger on it— is not as easily perceived.

In view of the difficulties of experimenting with the gravitational force, or changes in it, there is at present a better understanding and control of electric and magnetic forces than of the gravitational force. In addition, it has been difficult to relate the gravitational force to other phenomena. Thus, use of the electromagnetic field to control the gravitational force is not yet a feasible technology; indeed, such a use of electromagnetism may have to await the development of a theory which explicitly links known electromagnetic effects to the gravitational field, i.e. a successful unified field theory.

UNITARY AND UNIFIED FIELD THEORY

Attempts have been made to link, in theory, the electromagnetic and gravitational fields. The best known is the unitary field theory proposed by A. Einstein, based on formalistic (i.e. mathematical) considerations. In general, these considerations always include the notion that gravitation is associated with a geometrical property of space-time. This geometrical property, in turn, is justified by imagined measurements on the path of light signals in a vacuum. Since Einstein identifies the gravitational acceleration with the curvature in the paths traced out by the light rays, there is a very limited use of the properties of light. For example, those properties of light which belong to the subatomic level of observation are not called into play; consequently one may conjecture, on the basis of Einstein's own precepts, that those properties of the physical world on a subatomic scale are not adequately described in the unitary theory.

The theory upon which the following proposal is based, like the general theory of relativity and the unitary theory, depends on the properties of the electromagnetic field. However, the property featured in the proposed theory is not the effect of the gravitational force on the curvature of light rays, but rather the convertibility of matter into radiant energy and its converse. The latter is an experimentally observed phenomenon; the theory shows that it leads to a model, in terms of subatomic processes, for the generation of the gravitational force. Moreover, the model also suggests a method by which the gravitational force can be weakened or strengthened through the agency of electromagnetic interactions between subatomic processes. This method is suggested by an analogy between the cooling of a paramagnetic salt by use of a magnetic field—that is, a common laboratory technique—and the recognized method of dynamic nuclear orientation.

DYNAMIC NUCLEAR ORIENTATION AND GRAVITY REDUCTION

The basic mechanism proposed for reduction of the gravitational field is the "cooling" of the cloud of virtual particles/radiation surrounding a vehicle, and due to the elementary "virtual" particles and radiation surrounding the Earth. Since the gravitational field surrounding the vehicle is due to this cloud of virtual particles and radiation, reduction in the mean energy of the cloud is expected to result in the reduction of the gravitational force on the vehicle.

"Cooling," referred to above, takes place by first ordering the nuclear magnetic moments with respect to a fixed magnetic field. Since part of the magnetic moments of the nuclei is due to virtual processes, the virtual processes will also be ordered to that extent. The atoms of the structural material comprising the vehicle must be chosen so that the virtual particle cloud generated by the Earth (and which is disordered) has enough time to disrupt the ordered magnetic moments of the structure's nuclei, i.e. that part of the magnetic moment due to virtual processes. Moreover, the latter disruptions must occur before disruptive thermal processes intervene to ensure that it is the Earth's virtual particle field which is interacting with the nuclei, not local thermal processes. In the process of this disruption, it is expected that the virtual processes constituting the Earth's gravitational field will lose energy in the neighborhood of the vehicle, reducing the gravitational force on the vehicle.

Candidate materials for construction of a vehicle are governed by the observation that it is possible to orient the nuclei of a ferromagnetic element by the method of dynamic nuclear orientation. This method takes advantage of the fact that the electrons of these atoms are easily affected by a magnetic field. A constant magnetic field is first applied to the atoms of the material, causing the electrons to precess with a frequency proportional to the field strength. An oscillating magnetic field is then applied at right angles to the first magnetic field, with the same frequency as the precessional frequency of the electrons. The oscillating field need not be a strong field since, in general, a weak periodic disturbance can have a very large effect when applied to a physical system with the same natural frequency as the applied force. In the present case, the applied oscillating magnetic field tips over the precessing electrons' magnetic moments, thus aligning them with a far greater yield of common orientations than is possible with a much stronger constant magnetic field.

Moreover, since the electrons are very close to the nuclei of the atoms, the magnetic coupling between them acts to align the nuclei also.

It has been observed that it is possible to pass this nuclear orientation of ferromagnetic atoms to non-ferromagnetic atoms by immersing the ferromagnetic atoms in a matrix material such as aluminum. It is also found that at liquid helium temperatures, or near that temperature, the length of time the aluminum nuclei remain oriented is longer by a factor of about ten than the length of time the ferromagnetic nuclei remain oriented. Although the orientation lifetime becomes shorter, with respect to thermal processes, as the temperature of the surrounding medium rises, experimental data suggests that the lifetime at room temperature will remain long enough to serve the purpose of reducing the gravitational field.

CONSTRUCTION OF A VEHICLE

The combination of iron inclusions in aluminum is an especially attractive one for vehicle construction, for structural strength and ready availability of the metals. Although the resonance condition is difficult to maintain over a large spatial extent, one can imagine a small sample of iron immersed in a small sample of aluminum acting as a generator of the ordered state which is then conducted to the outer shell of the vehicle by means of aluminum cables. Once having reduced the gravitational force, other means of propulsion, such as Vernier rocket motors in space or in an atmosphere or propellers (in an atmosphere), can be used.

The advantages over rocket propulsion alone inherent in the proposed means of propulsion are:

a) A lower speed of takeoff from a planet, with consequent lower fuel consumption rate

b) A longer period of acceleration between planets, as well as a longer period of deceleration, leading to a shorter transit time

c) Vastly greater maneuverability

In addition, reaction mass can be collected along a given trip (e.g. if a nuclear reactor power plant is used) and extended trips of long duration become possible, without great dependence on home base for refueling or supplying reaction mass.

EXPERIMENTAL PROGRAM

As indicated in the above discussion, there are several areas which have not been verified by experiment. It is therefore necessary to institute an experimental program to eliminate these areas of ignorance before beginning on a vehicle construction program.

Most prominent of the unverified predictions is the relationship proposed between virtual processes and the gravitational force. Such experiments should first be performed at low temperatures to avoid interference from thermal effects. The dependence of the phenomena on temperature can then be determined, so that the feasibility of the proposed method of propulsion at elevated temperatures can be determined.

The most efficient combination of iron/aluminum, iron/magnesium, chromium/aluminum, and chromium/magnesium must be determined from theoretical analyses and laboratory tests.

Strengths of constant magnetic fields and microwave field should be determined for optimum performance, as well as the optimum microwave frequencies for given fixed magnetic field strengths.

Chapter 36

The UFT and the STR

INTRODUCTION (by D. Alzofon)

The unified field theory (UFT) is predicated on a minor correction to Einstein's special theory of relativity, a step that has engendered a bit of controversy over the years. "If it ain't broke, don't fix it," one physicist told me, and then, apparently believing he had dealt a crushing blow to the pretentions of "The Unity of Nature," my father's most comprehensive paper on the UFT, he handed it back and disdainfully refused to read it.

I asked my dad what he had to say about this criticism. The answer went something like this: "In exchange for a simple correction to the STR, I get unification of the fundamental forces, a solution to the wave/particle problem, elimination of infinities from field equations and quantum electrodynamics, and an operational definition of the gravitational force that yields engineering applications. Seems like a reasonable trade to me."

Misunderstandings about his motivations for modifying the STR arose often enough, however, to prompt him to write the following unpublished addendum to the 1981 paper. His 2001 paper, "Light Signals, the Special Theory of Relativity and Reality" may also have been written in response to this issue (see *Appendix C*, Ref. 7, p. 260).

The rest of this chapter was written by F. Alzofon.

INTRODUCTION (by F. Alzofon)

It is the purpose of this addendum to clarify the relation of the unified field theory (UFT) proposed in the paper "Anti-Gravity with Present Technology: Implementation and Theoretical Foundation" (AIAA-81-1608) to Einstein's special theory of relativity (STR, or *special theory*). It is proposed below that the UFT is a natural outgrowth of the special theory and is not a totally new beginning.

It is a consequence of this fact—the natural outgrowth of the UFT from the STR—that the UFT reduces to the special theory in the appropriate context, that is, it makes the same predictions as the special theory wherever the special theory is valid. However, the UFT extends the special theory to the region of high-energy processes where matter and radiation can be transformed into one another. Moreover, the UFT offers a means of removing troublesome and physically unrealistic infinities from the Coulomb-type laws that are valid for static fields (and low-energy processes), as well as those that occur in quantum electrodynamics. The latter features are not possible with the special theory in any direct manner.

Finally, the UFT provides a clear connection between the origin of inertial mass and the radiation field and a similar relation between inertial and gravitational masses. The special theory does not offer any such direct connections.

THE LIGHT SIGNAL

One of the central concepts introduced in the special theory is the notion of the light signal used for distant clock synchronization. There are, however, some assumptions made in the course of the clock synchronization procedure that must be altered in the light of present-day, improved knowledge of the properties of the electromagnetic field.

It is advisable, in this connection, to review the assumptions that lie at the root of Einstein's distant clock synchronization, since these are not always made explicit.

The use of the imaginary "observer" who is the agent for establishing a space-time coordinate system sometimes disguises that one is concerned here, not with a formal mathematical process, but with a real radiation field with experimentally determined properties. At minimum, this fact appears in the occurrence of the experimentally determined speed of light everywhere in the theory, even in the example used in Einstein's original paper on the special theory, in which the collision of two elastic spheres was invoked. That is, although the special theory was formulated to explain how material bodies interacted through the electromagnetic field, there was no mention of the electromagnetic field as a means of interaction in the latter example.

This feature could either be explained as due to the observer's use of the field for signaling, or to the fact that every material body interacts through the mediation of an electromagnetic field and in no other way. For example, in later developments of the special theory which se the Minkowski metric, it is pointed out that two events which cannot be connected by a light signal cannot affect one another at all, clearly a statement about the interaction of any two bodies. The latter is a statement about physical systems interacting, not about observers. Another example we can cite is the derivation of the Lorentz force from the Coulomb static electric field expression by use of the Lorentz transformation; the intervention of passive observers cannot be imagined to affect the accuracy of this description.

It is remarkable that only one experimentally determined parameter, the speed of light, is enough to yield all of the above-varied results. But it is as a result of an experimentally determined property of the electromagnetic field, and not of a mathematical concept alone that all these advantages are available. If there are other properties of the real radiation field which have a bearing on the interaction of material bodies, surely these ought to be included in any considerations leading to a theory describing these interactions. These properties must figure in every such interaction to be of value.

There is, however, more to the special theory than distant clock synchronization.

In the special theory, it is assumed that a light signal preserves its identity when observed by observers in relative uniform motion (i.e. moving with constant relative velocities). It is agreed that not all properties of the light signal remain invariant; for example, the frequencies (colors) composing the signal do change owing to the Doppler shift. But in terms of the idea of a light signal as a means of material body interaction there is a difficulty, for the interaction of two bodies may indeed depend on the frequency content of the mediating fields. Thus, as a physical entity, the light signal is, even in the context of the special theory, not an invariant.

Returning again to the special theory, the fact that the light signal retains its fundamental character to each observer (material system) as light, with a *characteristic speed of propagation*, is recognized by requirement that each observer that each observer sees the signal to be traveling with the same velocity, *c*, whatever their relative velocities. The consequent apparent contradictions in their interpretation of their measurements is resolved by the use of the Lorentz transformation: a dictionary which they can use to reconcile their differing sets of numbers used to describe the same physical event. More than this, the latter principle (of covariance) is extended to the assertion that the equations of motion for particles and radiation must have the same form in every observer's coordinate system.

Thus, by excluding all properties of matter and radiation except the speed of light alone, we can avoid the difficulty pointed out above in connection with the Doppler Effect. But where additional properties of the real radiation field might have a drastic effect on all interactions, *these properties cannot be ignored*.

MODERN VIEW OF THE RADIATION FIELD

Quantum electrodynamics has shown that the Maxwell electromagnetic field is not an adequate description of the real radiation field. In order to bring the predictions of theory into agreement with experiment, it was necessary to assume that the vacuum of classical theory was altered to contain an "infinity" of energy, with corresponding creations and annihilations of radiation and particles. This formal device had a precedent at the outset of the quantum theory in attempting to incorporate the possibility of the "spontaneous" radiative transitions which Einstein had found it necessary to introduce in his statistical analysis of black-body radiation.

Thus, a light signal passing through this infinite sea of energy could hardly avoid being altered in a fundamental manner: an alteration that would not become evident until a scale of measurement suitable to observe such effects was adopted. Such a scale of measurement was used in the observation of the Lamb-Retherford effect.

On the other hand, since Einstein has adopted the point of view that the vacuum possesses a structure determined by light signals used to probe it, this structure (or metric) must also be altered by the creation and annihilation processes introduced on the basis of accepted modern theory.

But it would be far more believable to ascribe the alteration in the interaction of material bodies, not to a vacuum occupied by an infinite amount of energy (a contradiction in terms), but to an intrinsic property of the radiation field. This is not without precedent: the radiation field is known to undergo fluctuations that, in principle, may become infinite, although with decreasing probability (on the basis of classical radiation theory).

It has been the approach of the UFT to incorporate all the above phenomena into the structure of space-time in a manner that familiar ideas and methods can be used as a practical matter of convenience.

SUMMARY

The above qualitative discussion has been offered as a means of justifying in more detail the development of the UFT in the referenced paper. It is emphasized that only well-established ideas and properties of the radiation field have been used; they have simply been combined in a different manner, e.g. the dispersion theory approach.

Chapter 37

A "New and Simple Idea"

INTRODUCTION (by D. Alzofon)

"A 'new and simple idea,' dark matter-energy and the crisis in physical theory" was my father's final paper on the UFT, written when he was 90 years old. It was dedicated to Richard Feynman, who had inspired him to complete his first paper on gravitation in 1954.

I remember his excitement when he completed the paper after months of work. In preparation, he had studied recent papers on general relativity in the library at Oregon State University. He realized he had difficulty in communicating his ideas to academics steeped in general relativity, and with this paper he felt he had finally found a way to reach them. He thought the paper had some extremely important insights and could not possibly be rejected by peer reviewers unless they were prepared to reject the vast body of accepted physical theory on which it was based. Nonetheless, it was turned down by *Nature* and *Physical Review* without comment. I remember him being bitterly disappointed by the unexplained rejections. There was no reading the minds of the reviewers, but he felt that all the old prejudices dating back to the 1940s must be in full sway in 2009. While there is objective evidence for the existence of these prejudices, this is not the proper place to elaborate. After thinking about it for a while, he decided that he didn't have the stomach for another time-consuming battle with editors and referees at other journals, so he filed the paper online at ViXra.org in 2009 in order to establish a copyright on the text and ideas. No permissions were required to reprint it here.

The following is a copy-and-paste of text from the ViXra.org website:

ViXra.org is an e-print archive set up as an alternative to the popular arXiv.org service owned by Cornell University. It has been founded by scientists who find they are unable to submit their articles to arXiv.org because of Cornell University's policy of endorsements and moderation designed to filter out e-prints that they consider inappropriate. ViXra is an open repository for new scientific articles. It does not endorse e-prints accepted on its website, neither does it review them against criteria such as correctness or author's credentials.

In order to download the original paper, go to http://vixra.org/pdf/1007.0008v1.pdf

A "NEW AND SIMPLE IDEA"

Dark Matter-Energy and the
Crisis in Physical Theory

ABSTRACT: *Correction of an omission in A. Einstein's operational definitions of time and space intervals in the special theory of relativity leads to an improved phenomenological and conceptual foundation for a previously proposed unified field theory. In combination with Einstein's researches on the fluctuation in energy of black body radiation, there results a "new and simple idea" of the kind Professor Richard Feynman felt to be necessary for the solution of the cosmological constant problem. A brief description of the formalism of the theory is presented. The infinite zero-point energy of the vacuum is eliminated. A model for the origin of inertial mass and dark matter-energy is deduced. The resulting relation between observed matter and dark matter-energy leads to a restriction on their magnitudes. The magnitudes of the latter quantities are then estimated from astronomical data. A model is proposed for the origin of the gravitational field in terms of a dynamic process at the basis of the proposed theory. The success of the special theory of relativity in predicting the results of three crucial observations establishing the validity of the general theory of relativity and the elimination of the infinite vacuum energy suggest that the unified field theory can lead to a solution of the cosmological constant problem.*

Key words: unified field, inertial mass, dark matter, gravity model, cosmological constant

Bracketed numbers are references to published sources (see p. 239).

1. INTRODUCTION

The late Professor Richard Feynman called for a "new and simple idea" [1] in connection with the cosmological constant problem [2]. The problem arises in the evaluation of a constant introduced into the equations of motion of the general theory of relativity (GTR) [3] and is related to the expansion of the universe [4].

Since the constant appears in a theory based on a model of the origin of the gravitation field, attempts at a solution of the problem necessarily involve phenomena on a cosmological scale. On the other hand, its relation to the expansion of the universe has brought into consideration the zero-point energy density of the vacuum [4], dark matter [2], and the nature of inertial mass [2], implicating subatomic phenomena.

Commenting on the connection between the latter topics, Professor Feynman pointed out that the zero-point energy in a vacuum would be expected to generate a gravitational field: instead "it is zero" [1]. As a consequence, Feynman suggested that the "new and simple idea" should also reformulate modern physics so that there is no zero-point energy in a vacuum [1]. Indeed, analysis of observations by the Hubble telescope has supported Professor Feynman's belief in the lack of any objective existence of zero-point vacuum energy [5, 6].

The need for a solution to the cosmological constant problem has been characterized as a "veritable crisis" in physical theory [7] whose solution may be expected to have a considerable effect on physics as well as astronomy.

This paper proposes a "new and simple idea," based on experiment, for the analysis of cosmological phenomena which meets Feynman's goal and is in accord with [5, 6]. The discussion emphasizes physical models, rather than a mathematical formalism, although an abbreviated mathematical basis for these models is provided. The theory has been discussed in more detail in several papers [8–10] for subatomic, atomic and terrestrial phenomena with a slightly different (but equivalent) basis, with the same formalism as presented below. It has been shown to lead to a unified field theory (UFT) of a matter-radiation field, reducing in the proper contexts to the Maxwell electromagnetic field, Newtonian and relativistic mechanics and quantum mechanics. A model for the origin of the gravitational field was also included. It was shown that the infinite zero-point vacuum energy can be eliminated, as well as other infinities which have long been characteristic of a zero separation of test body and classical field source.

2. POINT OF VIEW OF THE UNIFIED FIELD THEORY

The point of view in the formulation of the UFT is based on observation: a viewpoint diametrically opposed to the approach employed in Einstein's unified field theory [11]. In a general vein, Einstein states "…this axiomatic basis of theoretical physics cannot be extracted from experience but must be freely invented and can we ever hope to find the right way? …I am convinced that we can discover by means of purely mathematical constructions and the laws connecting them with each other, which furnish the key to understanding of natural phenomena …in a certain sense, therefore, I hold it true that pure thought can grasp reality, as the ancients dreamed" [12]. Further, Einstein's emphasis on mathematics and deduction as a source of inspiration for the creation of a new physical theory is summarized in his remark that "…the creative principle resides in mathematics" [12]. The latter point of view has permeated theories subsequent to the GTR, especially as they relate to a model for the gravitational force field. That is, the field is assumed to be identified with the curvature of the space-time metric—a geometrical property, rather than due to a physical dynamic process.

In contrast, we propose an alternative model for the origin of the gravitational force field resting on properties of subatomic phenomena, rather than on an intrinsic property of the space-time metric. This is expected to facilitate a unification of gravitation and the other three fundamental forces in nature—the electromagnetic, strong and weak forces. Further, the point of view adopted in the formulation of the UFT is in accord with a tenet advanced by H. Reichenbach, in his support of A. Einstein's original formulation of the special theory of relativity (STR) [13]. Based on a belief in logical positivism, it is asserted that, in advancing a physical theory, it is best to proceed by induction, in contrast to Einstein's dependence on deduction in [12].

H. Minkowski is more forceful in this regard: "My views of space and time have sprung from the soil of experimental physics and therein lies their strength." [14]

In addition, P. Bridgman states, in support of the STR [15], that every quantity introduced in a physical theory should have an operational definition, an opinion we shall implement below for the UFT. In this connection, we remark that such a requirement evidently can profoundly affect the interpretation of the data recorded, as, for example, the alteration of the length of a body when in motion relative to the observer.

Such an effect would not necessarily be observed were it not for the manner of measuring length as specified in the STR, i.e. by means of light signals [16].

The "simplicity" of the idea to be proposed lies in taking over the essential concepts and thought experiments of the STR. The "new" aspect of the idea refers to the alteration of the light signals used in the STR, in accord with the implications of [17] and [18]: that is, the fluctuations in intensity intrinsic to black body radiation are assumed to be a property of radiation propagated in a vacuum. The latter assumption is similar to that made by A. Einstein in [17] and [18] with respect to a quantum energy exchange property.

Moreover, to preserve logical and measurement consistency in the operational definition of a linear length in the STR, it is necessary to define coordinate magnitudes by means of light signals, i.e. for a zero, as well as nonzero relative velocities of coordinate systems.

In accord with [17 – 19], it is noted that the fluctuations in radiation intensity increase as the scale of measurements in space and time decreases.

The emphasis on the importance of radiation fluctuations is not only justified by [17 – 18], but also by researches in quantum mechanics, quantum electrodynamics and random electrodynamics [20]. It has also been shown that the formalism of nonrelativistic quantum mechanics is equivalent to classical mechanics on which is superimposed a random walk [21].

3. ALTERATION OF THE MINKOWSKI METRIC

As a consequence of the foregoing considerations, the average, observed, arrival time of a light signal t becomes $t - t_0$ where t_0 is a random variable such that $-\infty < t_0 < +\infty$ with a vanishing average value: $<t_0> = 0$. The nature of the statistical distributions to be used in this connection has been indicated elsewhere [8, 9].

Owing to the latter considerations, the equation descriptive of the propagation of a light signal along the x-axis of a rectangular Cartesian coordinate system becomes $x - x_0 = c(t - t_0)$ where x_0 varies over the same range as t_0 and with vanishing average value and c denotes the speed of light in a vacuum. Similar relations hold for the y and z coordinates.

When the above alterations are applied to the Minkowski metric of space-time

$$s^2 = x^2 + y^2 + z^2 - c^2 t^2 \tag{1}$$

and averaged, we find (where $r^2 = x^2 + y^2 + z^2$, for example)

$$<S^2> = r^2 - c^2 t^2 + <(r_0)^2> - c^2 <(t_0)^2> \tag{2}$$

The latter calculation suggests replacing the metric (1) by

$$S^2 = r^2 - c^2 t^2 + (r_0)^2 - c^2 (t_0)^2 \tag{3}$$

Setting $S = 0$ yields an equation descriptive of the propagation of a spherical light signal, subject to a fluctuation motion. Further physical interpretation of the metric (3) will be provided in Section 5.

The metric (3) is spatially flat (i.e. Euclidean) in agreement with observation [22]. In addition, the metric (3) is invariant under the Lorentz transformation:

$$x' = \gamma(x - \beta ct)$$

$$y' = y$$

$$z' = z$$

$$ct' = \gamma (ct - \beta x) \tag{4}$$

$$(x_0)' = \gamma(x_0 - \beta ct_0)$$

$$(y_0)' = y_0$$

$$(z_0)' = z_0$$

$$(ct_0)' = \gamma(ct_0 - \beta x_0)$$

where $\beta = v/c$, $\gamma = 1/\sqrt{(1 - \beta^2)}$ and v is equal to the relative velocity of two observers moving parallel to the x-axis.

Evidently the transformation (4), applied to the differences $X = x - x_0$, $Y = y - y_0$, $Z = z - z_0$ and $cT = c(t - t_0)$, results again in the invariance of the metric

$$S^2 = X^2 + Y^2 + Z^2 - c^2 T^2 \tag{5}$$

where X, Y, Z, and T are random variables such that $<X> = x$, $<Y> = y$, $<Z> = z$ and $<T> = t$. Subtracting the second set of four equations in (4) from the first set of four equations yields a Lorentz transformation for the quantities X, Y, Z, and T directly. Under the latter transformation the metric (5) is again invariant. In this case, however, X, Y, Z, and T contain "hidden" variables, and, together with the metric, can serve as part of a basis of a formalism for relativistic quantum mechanics. For the latter formalism, expectation values are calculated for a matter-energy (i.e. unified) field and not with the aid of a probability field which has no physical existence.

The differential form of the proposed metric (3) is

$$dS^2 = dr^2 - c^2 dt^2 + (dr_0)^2 - c^2 (dt_0)^2 \tag{6}$$

No quantization postulate has been introduced in the above discussion. This omission is supported by the derivation of the black body radiation energy spectrum without such a postulate; Lorentz invariance was, however, found to be essential [23].

4. MOMENTUM AND ENERGY OF THE UNIFIED FIELD

The following definitions have been chosen to parallel those of the STR [3]. In addition, it is found that the resulting formalism reduces to the STR in a suitable context.

Given some identifiable feature of the field described by the metric (6) at the point $(x, y, z, t, x_0, y_0, z_0, t_0)$, we define momentum components along the coordinate axes by

$$\boldsymbol{p} = m\,(dr/d\tau) \text{ and } \boldsymbol{p}_0 = m(dr_0/d\tau) \tag{7}$$

where $r = (x, y, z)$, $r_0 = (x_0, y_0, z_0)$, $d\tau = +\sqrt{(dt^2 - dr^2/c^2)}$, and energies

$$E = mc^2\,(dt/d\tau) \text{ and } E_0 = mc^2\,(dt_0/d\tau) \tag{8}$$

and where m is a factor with the dimensions of mass, whose nature will be clarified below.

Inserting the condition that neighboring observers can communicate with one another, i.e. $dS = 0$, into equation (6) and dividing the resulting equation by $d\tau^2$, we find

$$p^2 + (p_0)^2 = (E/c)^2 + (E_0/c)^2 \tag{9}$$

If the average displacement dr of the given feature be set equal to zero, then $d\tau = dt$ and

$$E = mc^2 \tag{10}$$

and

$$(p_0)^2 - (E_0/c)^2 = (mc)^2 \tag{11}$$

In view of equations (10) and (11), we interpret m as the rest mass of the given feature of the UFT field, generated by the fluctuating motion of the field. This motion is proposed to be the origin of inertial mass. Averaging preserves the form of equation (11).

In the opposite extreme, we set $dr = c\,dt$, implying that $dr_0 = c\,dt_0$. This state of motion is readily interpreted as a light signal accompanied by fluctuations propagated with the speed of light c.

In view of the properties of the extreme conditions of motion, indicating a duality in the nature of the UFT field, we propose that matter be denoted as "condensed radiation" and radiation be denoted as "dispersed matter."

At intermediate average speeds, the given feature of the field can be expected to have both matter and radiation properties.

In view of the latter mass and radiation properties of the field introduced, we call it the "matter-radiation field." Its ability to incorporate many other fields in its formal structure [8–10] leads to its designation as a "unified" field.

It is noted that the concept of a spherical light signal of the STR has been replaced by a spherical signal which includes matter states as well as radiation states; that is, two events may be connected by matter as well as radiation properties. In contrast, the spherical light signal of the STR refers solely to radiation.

The model corresponding to the above considerations implies that all observed matter and radiation is accompanied by fluctuations in energy with matter and radiation properties.

There is evidently a close analogy between the latter model and the concept of vacuum energy. We therefore propose that the vacuum zero-point energy concept be replaced by the UFT model, which is Lorentz invariant and yields finite results; it is not an intrinsic property of space-time, in accord with observation [5, 6]. Moreover, the useful deductions sometimes held to be evidence for the objective existence of the zero-point vacuum energy (e.g. the Casimir Effect) can be preserved with the latter revision, since it is acknowledged that their explanation relies on energy fluctuations, rather than on the background energy [24].

5. DARK MATTER-ENERGY

With reference to the remarks of the preceding section, it follows that all matter observed by optical and radio telescopes has additional and unobserved inertial mass characterized by very high frequency motion. It can, however, be detected by its effect on the angular momentum of large amounts of observed matter and by its gravitational effects.

It is proposed that the additional random motion accompanying radiation and matter observed by optical and radio frequencies is responsible for the phenomena associated with dark matter.

[*Ed note: The previous two paragraphs appeared in a draft version of the paper, but not in the final. They were restored here because it seemed possible that they were omitted by mistake. A handwritten note, "Check discovery of dark matter," appeared in the margin. That check may have led to their being deleted.*]

Since the averaged, observed, motion of matter and energy, characterized by the four-vector (p, E/c), is accompanied by a fluctuating motion described by the four-vector (p_0, E_0/c), giving rise to inertial mass-energy (see eq. 11), we propose that dark matter and energy are root-mean-square consequences of the latter motion of the unified field.

In this connection, Professor Steven Weinberg has called for an explanation of why dark energy (here $\sqrt{[<(E_0)^2>]}$ is the same order of magnitude as energy in observed matter, i.e. $E = mc^2$ [25], implying that each acts as a constraint on the other.

The latter approximate equality becomes plausible in view of equations (10) and (11) and the proposed origin of dark matter-energy. Since Professor Weinberg specifies more closely that $\sqrt{<(E_0)^2>}$ is of the order of $2mc^2$, we deduce from (11) that $\sqrt{<(p_0)^2>}$ is approximately equal to $(\sqrt{5})(\sqrt{[<(E_0)^2>]}/c)$. this relation is to be compared with the condition for observing radiation alone for which the factor $\sqrt{5}$ (approximately 2.24) is replaced by unity. These relations, then, may be an aid in formulating an explanation for the observed magnitudes of dark energy and matter and illustrate the usefulness of the "new and simple idea."

6. A MODEL FOR THE ORIGIN OF THE GRAVITATIONAL FIELD

The following discussion proposes a mechanism for the origin of the gravitational field based on the UFT's inertial mass model and the principle of Le Châtelier [26], which states, essentially, that if a small perturbation be applied to a system in equilibrium, its parameters change in such a way as to restore equilibrium. It offers the possibility of removing the specialized role of the gravitational force in the GTR (i.e. a geometrical model, identified with the curvature of space-time), while the other forces in nature are held to be due to physical processes. The latter disparity hinders a unified representation of all these fields.

The Newtonian gravitation field exists in the presence of matter and not otherwise. It has been observed that all matter is composed of subatomic particles and therefore gravitation is a property common to and originating in these. Consistent with equations (10) and (11), we view these particles as bound states of the matter-radiation field with an internal motion described by equation (11), and with most of its mass m confined to a spherical region with diameter h/mc, where h denotes Planck's constant. It is assumed that these particles are stable during the period of observation in accord with a similar assumption about the macroscopic state of matter in Newton's law of gravitation. In effect, then, one is dealing with a physical system with internal fluctuations in energy analogous to a gas at a uniform temperature, composed of randomly moving molecules.

These concepts are the basis of the following approximate analysis.

Consider first a single mass particle-field of mass m. When a second particle-field is brought into the neighborhood of the first, the inertial mass of the first field increases to $m + \Delta m$ (we assume that $\Delta m/m$ is very small compared with unity) so that, by Le Châtelier's principle, the new equilibrium diameter of the field becomes essentially $h/(m + \Delta m)c$, or approximately $(h/mc)(1 - \Delta m/m)$ and mass energy (including that from the second particle-field) will tend to flow toward the center of the first particle-field.

Assuming that each particle-field retains its identity, there will be generated a force tending to move the second particle towards the first. A similar argument applies to the second particle-field so that there is a mutually attractive force urging the two particles together.

The equations of motion of the UFT for a point source of a static field (i.e. assuming a Newtonian type of field) requires that the field potential be proportional to $1/r$, where r represents the distance between source and test body.

We are then led to the plausible assumption that the force exerted by a subatomic particle on another subatomic particle by reason of its internal random motion, and at a distance greater than a Compton wavelength, can be identified with the Newtonian law of gravitation.

7. THE COSMOLOGICAL CONSTANT PROBLEM

The severity of the cosmological constant problem becomes apparent upon comparison of estimates of the constant from cosmology, limiting its absolute value to less than $10E^{-56} cm^{-2}$, while estimates based on zero-point energy (i.e. modern particle physics) differ from this limit by forty orders of magnitude [27]. Further evidence of the deep division between the two theories lies in Professor Feynman's remark that zero-point energy has no gravitational effect (and therefore no point of contact with the GTR), while the GTR has little to say about modern particle physics. At minimum, then, it would appear that a unifying physical model which leads to a finite replacement for zero-point vacuum energy, as well as a model for the origin of the gravitational field, would be a promising candidate for resolving the cosmological constant problem. A further complication to be resolved is that particle theory is often discussed in terms of the STR with an implicit assumption that all coordinate reference frames are in uniform relative motion, while the GTR deals with accelerated frames of reference.

To be sure, for sufficiently small regions of space-time, the GTR can be approximated by the STR [28]. Moreover, it has been shown that calculations establishing the validity of the GTR for three crucial observations can be replaced by estimates based on the STR.

These are: [29], for the red shift in radiation emitted by atoms in a strong gravitational field and the deflection of a light ray passing the sun, as well as [30], for the magnitude of precession of the perihelion of Mercury's orbit.

As a consequence, it would appear that if the UFT be added to the capabilities of the STR, with a consequent unification of particle and gravitational field theory, there could result an improved estimate of the cosmological constant, consistent with both theories.

8. SUMMARY

The preceding discussion has had two principal objectives:

The first objective was to show that, since the special theory of relativity is based on operational definitions using clocks and realistic light signals, the definition of space-time coordinate intervals must include the effect of light signal energy fluctuations on their magnitudes. This requirement is a *necessity* for the logical consistency of any resulting theory and leads in a natural way to a unified field theory which contains previously derived theories and their useful consequences.

The second objective was to show how the referenced unified field theory, when applied to cosmology, can serve as the "new and simple idea" felt to the necessary by Professor Richard Feynman for solution of the cosmological constant problem. As demonstrated above, the theory meets Feynman's principal requirement that the infinite energy density of the vacuum be eliminated. From the expressions derived, it is then shown that there follows an explanation for the origin of dark matter and why it is of the same order of magnitude in energy as observed matter: an answer to a question posed by Professor Steven Weinberg.

Since the unified field theory lacks the infinities associated with the special theory of relativity, it is expected that it can aid in solution of the cosmological constant problem, especially since the fluctuations in matter-energy can be linked to observation through equation (11).

REFERENCES (for "A New and Simple Idea")

1. P. Davis and J. Brown, *Superstrings, A Theory of Everything* (Cambridge University Press, New York, 1988)

2. A.H. Guth, *The Inflationary Universe* (Addison-Wesley, New York, 1942)

3. P.G. Bergman, *An Introduction to the Theory of Relativity* (Prentice-Hall, Inc., New York, 1948)

4. M. Roos, *Introduction to Cosmology*, 2nd Ed. (John Wiley & Sons, New York, 1997)

5. R. Ragazzoni, M. Turato and W. Gaessler, *Astrophys. J.* **587**, L1 (2003)

6. R.R. Britt, http://www.space.com/scienceastronomy/quantum_hits_030402.htm

7. S. Weinberg, *Rev. Mod. Phys.* **61**, 1 (1989)

8. F. Alzofon in AIAA/SAE/ASME 17th Joint Propulsion Conference, 1981, Colorado Springs, AIAA-81-1608

9. F. Alzofon, *Physics Essays* **6**, 599 (1993)

10. F. Alzofon, *Physics Essays* **14**, 144 (2001)

11. A. Einstein, *The Meaning of Relativity,* 3rd Ed. (Princeton University Press, Princeton, N.J., 1950)

12. A. Einstein, in *Essays in Science* (The Wisdom Library, The Philosophical Library, New York, 1934)

13. H. Reichenbach, in *Albert Einstein, Philosopher Scientist,* Ed. P. Schilpp, (Open Court, La Salle, IL, 1970)

14. H. Minkowski, in *The Principle of Relativity* (Dover, New York, 1952)

15. P.W. Bridgman, in *Albert Einstein, Philosopher Scientist*, Ed. P. Schilpp (Open Court, 1969)

16. A. Einstein, in *The Principle of Relativity* (Dover, New York, 1952)

17. A. Einstein, *Phys. Zeits.* **10**, 185 (1909)

18. A. Einstein, *Phys. Zeits.* **10**. 817 (1909)

19. L.D. Landau and E.M. Lifschitz, *Statistical Physics*, 3rd Ed., Part 1, Course of Theoretical Physics, Vol. 5 (Pergamon Press, New York, 1982)

20. T.H. Boyer, *Phys. Rev.* **11**, 790 (1975)

21. E. Nelson, *Phys. Rev.* **150**, 1079 (1969)

22. P. de Bernardis, P.A.R. Ade, et al. *Nature* **404**, 955 (2000)

23. T.H. Boyer, *Phys. Rev.* **150**, 1374 (1969)

24. T. Padhmanabhan, *Class. Quant. Grav.* **22**, L107 (2005)

25. S. Weinberg, http://supernova.lbl.gov/cylinder/weinberg.ps

26. *McGraw-Hill Dictionary of Physics and Mathematics*, Ed. D.N. Lapedes (McGraw-Hill Book Co., New York, 1978)

27. S.E. Rugh and H. Zinkernagel, in *History and Philosophy of Science*, Part B, **33**, 663 (2002).

28. W. Pauli, *Theory of Relativity* (Pergamon Press, New York, 1958)

29. L.I. Schiff, *Am. J. Phys.* **28**, 340 (1960)

30. R.J. Buenker, *Apeiron* **15**, 509 (2008)

PACS (Physics and Astronomy Classification Scheme), Nos. 03.65Pm, 03.65Ta, 03.65Ud, 11.10Kk, 98.80Jk

Chapter 38

A Word About Expert Opinion

"Science is the belief in the ignorance of experts."

— *Richard Feynman*

THE DISCLAIMER (p. 5) recommends submitting the book to an expert for review before attempting to duplicate the 1994 experiment, or for that matter, embarking on a gravity control R&D program.[68] If the reader heeds this advice, then the material in *Book II* will be the primary focus for expert review, though any comprehensive evaluation should include the 1981 Colorado Springs paper as well as the 2003 paper, *The Unity of Nature and the Search for a Unified Field* (see Ref. 4, p. 260). The pitfall for readers genuinely in a search for objectivity is that academic specialists in gravitation are the *least* likely experts in the world to render an objective opinion.

While a complete analysis of this peculiar state of affairs would require a book in itself, this chapter will explain it to some degree, without marshaling all the examples of less-than-objective treatment or conjecturing too deeply about the motives of the individuals involved. It will also go some distance toward answering a question that has no doubt occurred to many readers: "If this theory and the technology are so great, and they've been around for more than thirty years, how come somebody hasn't done something about it already?" *The History of an Idea* (p. 57) will provide additional insights.

Let's begin with a direct answer to the question, "Who's an expert?" That is, who do *we* consider most likely to render an informed, unbiased opinion of the technology and its theoretical foundation?

The answer: PhDs in electrical engineering, particularly those who specialize in microwave technology or electron paramagnetic resonance (EPR), a technique used in electron *spin* resonance (ESR) spectroscopy. Particle physicists are another possibility. A stronger than usual background in mathematics will be required in either case. An understanding of special relativity is highly desirable, though strictly speaking, not necessary.

And who is *unlikely* to render an unbiased opinion? Ironically, it is academic specialists in Einstein's general theory of relativity (GTR), which is the gold standard for gravitation theories, with every other

---

[68] See p. 1

scientific theory running a distant second or limping along in the crackpot leagues.

Before we begin, a word about "theory" itself. A scientific theory is not the same as, say, a "theory" that the Moon is made of green cheese or that "the universe is a living organism." A scientific theory has rigorous standards. Among other things, it must incorporate previously established facts and lead to predictions, *numerical* predictions. Predictions made by a scientific theory are not the same as predictions made by seat-of-the-pants guesswork, dreams, visions, or tea leaves. For more on what goes into a scientific theory, see page 181. The general theory of relativity meets all the criteria of a scientific theory. So does the UFT, although both theories deal with the same phenomena. They are therefore *competing* explanations of reality. Comparisons between them will emerge in the discussion below.

Before getting into the weeds, however, I want to say that there has never been any sort of a "conspiracy" to suppress my father's work. All of the physics professors I've met—and there were many of them in Palo Alto, Stanford, Santa Barbara, and Berkeley over the years—seemed to be good people dedicated to pursuing truth within their area of expertise. But "truth" was narrowly defined by a set of assumptions, an understanding of which required *years* of dedicated study. The general theory of relativity was one such bulwark. My father challenged it, albeit indirectly, and he did so as an outsider, since he had left academia for aerospace in 1956. This, in my opinion, accounted for their resistance.

As the stature of the GTR has grown, so too has resistance to Alzofon's unified field theory (UFT). The UFT's point of departure is Einstein's *special theory of relativity* (STR). The STR preceded the GTR, and while the two *are* related, the STR is not regarded by GTR specialists as a gateway to understanding gravity, much less a unified field theory. While the issue could be settled—and *was* settled to my father's satisfaction in 1994—by *one simple experiment*, academics have never shown any interest in understanding the UFT, let alone putting it to the test.

Does this mean the UFT is fatally flawed or beneath consideration? The results of the 1994 experiment *alone* would argue otherwise, but let's examine academic response from 1960 to 2012: In all that time, no academic ever argued against the UFT on its own merits. Instead, they reacted one of three ways: they ignored it, they raised irrelevant objections, or they reacted emotionally and refused to talk about it, such as the professor who hissed, "How dare you even *speak* of such things!" and stalked away with elbows pumping (not an uncommon reaction). In short, they have done backflips in order to avoid studying it or confronting it. The few who did debate my father learned that he understood the assumptions behind *their* position, i.e. the assumptions behind the GTR, better than they did (see p. 183), and they did not fare too well. My father used to say that academics preferred to ignore the UFT because they couldn't reject it without rejecting a body of facts at the core of modern physics.

What facts? Let's start with Einstein's special theory of relativity, which is the springboard for the UFT, as it was for the GTR. The STR is one of the most thoroughly validated scientific theories in history. The premise of the UFT—the substitution of a realistic, fluctuating light signal for the ideal light signal of the thought experiment—is based on experimentally verified fact, not idle conjecture. The reasoning constructed on this slightly modified foundation is in the classical tradition that served physics so well for hundreds of years. My father argued that physics went astray when theoreticians chose faulty solutions to tricky problems that emerged in the late nineteenth and early twentieth century. This isn't any kind of revelation; the problems have been known for a long time and are readily acknowledged. It's just that the difficulties and their solution are now viewed as integral contradictions in nature itself. But the paradoxes and strangeness they produce have inaugurated a strain of mystical thinking in physics, which has been reinforced in popular literature. This provides more than ample justification for revisiting the STR. Quoting my father:

> *As a consequence of the natural outgrowth of the UFT from the STR, the UFT reduces to the special theory in the appropriate context, that is, it makes the same predictions as the special theory wherever the special theory is valid. However, the UFT extends the special theory to the region of high-energy processes where matter and radiation can be transformed into one another. Moreover, the UFT offers a means of removing troublesome and physically unrealistic infinities from the Coulomb-type laws that are valid for static fields (and low-energy processes), as well as those that occur in quantum electrodynamics. The latter features are not possible with the special theory in any direct manner. The UFT includes, as special cases, Newtonian mechanics, relativistic mechanics of a mass particle, Maxwell's equations of the propagation of the electromagnetic field, the gravitational field, and a continuous gradation of these into forces which may be identified with nuclear forces on a sufficiently small scale. It also makes predictions identical to general relativity. Consequently, it may be said that there is already a considerable body of experimental verification available for the UFT, even without the 1994 experiment.*

> *Finally, the UFT provides a clear connection between the origin of inertial mass and the radiation field and a similar relation between inertial and gravitational masses. The special theory does not offer any such direct connections.*

This is a rebuttal to a physics professor who, like so many others, didn't feel he needed to read "The Unity of Nature," because, as he put it (referring to the STR), "If it ain't broke don't fix it." That old time religion was good enough for him, in other words.

As noted by Dr. Hal Puthoff, Stanford Physics professor, a professional associate of my father's at SRI in the 1950s, and friend through the 1990s, the GTR explains gravitation via a geometric construct and does *not* provide any understanding of gravitation at a fundamental level.[69] Quoting from one of my father's emails:

---

[69] See H. E. Puthoff, "Gravity as a Zero-Point Fluctuation Force," Phys. Rev. A 39 (1989) 2333.

The reasoning for Einstein's remark that the metric of space-time is a "real" entity is that the gravitational field is part of it and is observable, and therefore the metric is real. I don't believe in this approach, of course.

And, quoting from his final paper on the UFT (p. 229):

The latter point of view has permeated theories subsequent to the GTR, especially as they relate to a model for the gravitational force field. That is, the field is assumed to be identified with the curvature of the space-time metric—a geometrical property, rather than one due to a physical dynamic process. In contrast, we propose an alternative model for the origin of the gravitational force field resting on properties of subatomic phenomena, rather than on an intrinsic property of the space-time metric.

In 1913, Max Planck said that gravitational theories were "falling thick as hail," making it hard for him to know where he might find a kernel of truth.[70] Planck might well have been interested in the UFT, however, because of its amenability to experimental verification, its ability to make many of the same predictions as the GTR, and its unique prediction that the gravitational force can be manipulated with present technology. There is *no* room for such a prediction in the GTR.

To summarize, the UFT is *not* radical in spirit, nor is it a departure from the historical evolution of physical theory. Rather, it is well-integrated into physical theory up until the time of the GTR. Like the GTR, it used the STR as a starting point, but it went another direction by adding a more realistic light signal to the STR. The UFT resulted from the inventor's deep study of particle physics and relativity in graduate school at UCLA and Cal Berkeley in the 1940s and 1950s. He was well aware of all the objections which could be raised against it, because he argued these with some of the best physics professors in the world, such as relativity expert Victor Lenzen and Richard Feynman, either of whom would have been delighted to punch a hole in the UFT. They failed.

At the very least, the UFT's respectable lineage and experimentally verifiable (we would say in all probability *already verified*) predictions make it worthy of discussion. In spite of that, academic response to the UFT has been oddly emotional and evasive, which suggests some underlying fear or prejudice. Examples are unfortunately abundant (a few have already been cited in the text), but one needn't look far for a motive, since the UFT completely *sidesteps* the GTR. If it is proven correct, it will, like Copernicus's heliocentric theory of the solar system, topple the edifice on which hang lifetimes of study, careers, grants, salaries, honors, publication contracts and royalties—not to mention the effect it would have on the self-image of many brilliant scientists. Everything considered "real"

---

[70] *Intellectual Mastery of Nature, Theoretical Physics from Ohm to Einstein*, Vol. 2, by Christa Jungnickel, Russell McCormmach, University of Chicago Press, 1990, p. 329.

within this world would suddenly be up for grabs, a most uncomfortable proposition.

As Professor Thomas Kuhn has shown,[71] social factors in science are never mentioned overtly, but to pretend that they don't exist would be the height of absurdity. Experience has shown that they cast a rather long shadow over academic thinking, wherein the power to control high-level discussion on gravitation is vested, and of course, the ability to control funding for gravitation research.

This book is not intended as a critique of general relativity. Nor is it intended to be an attack on a certain segment of academia or a defense against criticism from that quarter—especially since there *isn't* any criticism to defend against. Indeed, the thundering silence of academia is one of the stranger aspects of this narrative, since the inventor published many peer-reviewed papers and books in his lifetime, including a paper on "Relativistic Neutron-Proton Scattering in the Born Approximation," in *Physical Review*, and was an acknowledged world-class expert in two areas where relativity plays a role: optics and radiation scattering from objects of arbitrary shape. In short, he knew whereof he spoke. Only in the area of gravitation and unified field theory did the "reality distortion field" take hold.[72] The purpose of this chapter is to answer the question "Who's an expert?" and justify our answer by establishing the integrity of the UFT and explaining how it is perceived by experts in the GTR, the only theory of gravitation with any currency these days.

Whenever my father submitted a paper on the UFT to the major journals, he used to brace himself for the peer reviewers to discover the fatal flaw he had overlooked. What he got—when he got anything at all—was irrelevant nonsense, if not hysterical ranting. I read some of the comments myself, and even though I am not a physicist, it was easy to see that not one reviewer *ever* addressed the central points of his thesis. Indeed, it seemed as if they were purposely avoiding confronting it head-on, or worse, they had never even read the paper at all. Since the major journals have no "court of appeal," no way to respond to peer-review criticism, he was forced to turn to a second-tier journal that did allow it. The peer review criticism there exhibited the *same* characteristics it had in the major journals. His rebuttals to the reviewers were rather devastating, which is why the editor saw fit to publish.

Academics who were accustomed to thinking of themselves as the smartest guys in the room all their lives had a tendency to underestimate my father for a couple of reasons:

First, he left academia for aerospace in the mid-1950s. To them, this seemed to mean that he had no authority to speak (or publish) on matters relating to theoretical physics, that is, no standing to speak

---

[71] Kuhn, Thomas, *The Structure of Scientific Revolutions*, University of Chicago Press, 1962 (50th Anniversary Edition, 2012), ISBN: 9780226458113

[72] Bud Tribble coined this useful term to describe the effect of Steve Jobs' charisma on his surroundings. I heard it for the first time while working at Information Appliance in the 1980s, but not from Jef Raskin or Bud Tribble (who worked there briefly). Rather, it was one of the IAI engineers who had worked at Apple around the same time as Jef.

about the GTR, where all questions about gravitation are referred first.

Second, they underestimated the depth of his study of relativity or his knowledge and experience in experimental science. They did not know of his habit of questioning every assumption that his professors (and theirs) glossed over on the path to the GTR. One of the recipients of his relentless questioning was Professor Victor Lenzen, a leading authority on relativity and a student of logical positivism under Bertrand Russell. Lenzen was quite happy to debate the fundamental assumptions of the STR and the GTR, or any aspect of modern physics, with my father. It was a good training ground.

When professors began to throw mathematics at my dad, they came up against more unexpected depth. The source was his mentor and PhD advisor, Professor Griffith C. Evans, namesake of Evans Hall of Mathematics at UC Berkeley. (My father revered Evans more than any professor he had at Cal.) The professors who debated the UFT with him found the experience disconcerting. The nobody with the eccentric notions about relativity became a threat in short order, as he made them look like they didn't know what they were talking about. Most of his opponents would assert their superior status, usually with an outburst of righteous indignation, and then run away as fast as possible.

While my father was an expert in relativity, particle physics, applied mathematics, and the scientific method by the time he left graduate school, he continued to supplement his knowledge by reading current and past publications, especially on relativity—many in the original German or translated from Russian—all the way through his 93rd year. I never saw him spend fewer than six hours a day reading and writing about physics. He always knew that there was a gulf between his thinking and that of experts in general relativity, and he wanted to find better ways to communicate with them, so he read all the latest material on general relativity at the OSU library. He found nothing there to persuade him that he was on the wrong track. On the contrary, the more he read, the more convinced he became that he was right. He also became more pessimistic about persuading anyone who was committed to general relativity that another way was possible.

Unlike Nikola Tesla, however, he had no ax to grind with Albert Einstein. Indeed, he revered Einstein more than any other physicist and would have been the first to acknowledge that the UFT was merely a modest extension of the STR. The proof can be found in a 1955 letter he wrote to Einstein expressing his admiration for the latter's work and describing an early form of the UFT (see p. 63). Unfortunately, Einstein died April 18th, 1955, before any dialog was initiated. The letter to Einstein had been prompted by his 1954 meeting with Richard Feynman, who told him he was "on to something" and encouraged him to publish (see p. 59). The meeting with Feynman, more than anything, set in motion the events described in this book. Feynman was the dedicatee of my father's final paper on the UFT in 2009 (see p. 229).

In short, the UFT is not an amateurish construct. Rather, it is soundly and thoroughly based on a slight, well-justified modification to Einstein's special theory of relativity. From this modest "correction," he derived an operational definition of gravitation and many other significant results:

The success of the special theory of relativity in predicting the results of three crucial observations establishing the validity of the general theory of relativity and the elimination of the infinite vacuum energy suggest that the unified field theory [UFT] *can lead to a solution of the cosmological constant problem.* (See p. 229).

We have stated that this book is not an attack on the GTR. Nor are we here to defend the UFT. It is enough to establish the intellectual pedigree of the UFT and use it to posit an experimentally verifiable explanation of the origin of the gravitational force, one that has been formulated according to the rules of the scientific method, and that—unlike the GTR—can readily be harnessed to applied technology, such as the revolutionary propulsion system described in this book. That is all we intend to do, and if organized science were as true to its ideals as it advertises to the world, the theory would have been tested long, long ago. Instead, it has been ignored, with potentially devastating consequences.

If the reader agrees that the UFT is worthy of investigation, and contrary to our advice submits this book to an "expert in gravitation," the reaction will be fairly predictable: They will not be intrigued by the theory nor interested in its predictions. Rather, they will first glance at where the UFT was published, and then dismiss it immediately when they find out it was not published in a top-tier journal. If they read further at all, they will view it as an assault on all they hold sacred, even though it scarcely mentions general relativity at all. You may expect the criticism to begin early, and to become quite heated. Rejection, if not ridicule and contempt, is a certainty. If on the other hand, you give this book to a PhD in electrical engineering, they will grasp its points immediately and they will see that it refers to established knowledge and known techniques, not guesswork and humbug.

If you seek the opinion of an expert in the GTR, bear in mind that academia has long acknowledged that there are difficulties with general relativity. At the same time, they have consistently sought to fix these difficulties by tweaking the GTR, *never* by reconsidering its basic assumptions. This is even less likely now than in 1960, when Dr. Maurice Garbell, author of an Air Force-commissioned worldwide survey of gravitation research (see *Appendix C*, p. 259, Ref. 1a), told my father that the UFT would never be investigated at *any* academic institution, even though it was the only theory on gravitation that Garbell had seen that "had a prayer of an engineering application."

The possibility of settling longstanding issues in the GTR, the allure of an easy Nobel Prize, and the exciting prospect of opening a new frontier in technology and space travel—all of this has held little romance for academia, which has steadfastly resisted even *looking* at the UFT. Meanwhile, they have launched big-budget, empire-building projects to explore general relativity and particle collisions. The effort to detect gravity waves is a case in point (more below).

While the GTR has made many good predictions, the simple fact that unlike, say, Maxwell's equations, it has not resulted in a *single instance* of applied technology somehow gets lost. This departure from classical physics, which has always resulted in an abundance of applied technology, does not strike its partisans as the least bit symptomatic of a fundamental conceptual flaw. Rather, the absence of progress in applied technology seems to them to represent an insurmountable wall built into the fabric of nature, rather than an illusory barrier built into the assumptions of the GTR. To suggest breaching that wall, or even to suggest that such a thing is possible, provokes patronizing smirks, indignation, and in some cases, wrathful condemnation, and *always* an abrupt end to the conversation.

This brings up the topic of gravity waves, which were predicted by the GTR at the beginning of the twentieth century and confirmed in 2016. My father and I discussed this subject a few times over the years, and he said that whether or not gravity waves were detected, it would have no impact on his theory of the origin of gravitation. The *converse*, however, may *not* be true. It is my guess (not his) that if the 1994 experiment is repeated successfully, it will force a reinterpretation of the results of 2016.

It is regrettable that this subject needs to be brought up at all, but the academic blockade of my father's gravitation theory has been an ongoing and at times weirdly comical sideshow in this saga for more than fifty years. If it were *not* mentioned, it would tend to have a corrosive effect on the rest of the narrative. In spite of the extended airplay I've given it here, it really does not deserve the spotlight. If my father is *correct*, the resistance of academia is something for a scientific historian such as Thomas Kuhn to sort out, and if the UFT is *not correct*, any brouhaha would be moot.

Make no mistake: this book is *not* a plea for vindication of the UFT. Professors can hurl stones at the points raised here, but their judgment is completely *irrelevant*, since we are not taking our plea before a court of academics. We are taking it to the ultimate court of appeal, the court of Nature. Fifty years ago, the verdict might have been a matter of academic interest only. Today, it might well determine the future of humanity. If there's anything to it, that is. We won't know that until someone redoes the experiment.

Since academics are likely to take umbrage at the foregoing characterization, let me say that this is merely the account of an eyewitness to fifty years or so of history. I am not qualified to argue the virtues or faults of the GTR vs. the UFT, and the time when physicists could have debated the question with my father has come and gone. Let me reassure any insulted parties that his criticisms of the ivory tower were *extremely* mild compared to what they might have been. I would describe him as saddened, rather than angered, by the reception given the UFT, and the thing that saddened him most was not the lack of recognition—it was the catastrophic loss of time.

In 1981, his invention might have averted climate change entirely while opening the space frontier and stimulating the global economy to an extent not seen since the Industrial Revolution. By 2012, he thought that its only use might be as a lifeboat for a few survivors of the coming catastrophe, and that this group would most likely be self-selected from among the upper one-tenth of one percent of the economic strata. I don't know whether I agree with this Morlocks-and-Eloi vision of the future, but I do know that he never wanted gravity control to belong to the military and the wealthy elite alone. This is one of the reasons I have emphasized historical parallels to the beginning of the first Industrial Revolution. This isn't science fiction and there are no secrets about it. It is simply another step in the evolution of technology, and like almost all the others, it is merely the application of two old, well-understood technologies in a new way.[73]

In my view, trusted institutions have failed their mission, which is, nominally at least, to further the cause of science and humanity. Occurrences like this are rare, but not without precedent. On those occasions when they do occur, it becomes necessary to seek alternate routes to advance a legitimate cause, even routes that are unconventional, such as the present volume. As an institution of one, I don't relish the idea of riding into the public arena on a lone horse, waving a rebel flag under the noses of academics and other powerful interests, but given the stakes, it is a dire necessity—for *everyone*.

To summarize:

Not all experts are created equal. The experts most likely to read the book with an open mind are PhDs in electrical engineering. The least likely to render an objective verdict are experts in gravitation who have made a lifetime study of the GTR.

Who is right and who is wrong cannot be settled by scrutinizing the pedigree of the UFT or dueling with chalk sticks at a blackboard. No criticism from *any* quarter can be taken seriously as long as the critics are unwilling to rerun the 1994 experiment. If and when anyone decides to take up the challenge, we are confident that they will act honestly and honorably and

- Follow instructions carefully; making modifications *only* when absolutely justified
- Give a fair and accurate accounting of the results
- Duly credit the proper source

Consult *Appendix F* (p. 269) before filing patents

---

[73] Referring to adiabatic demagnetization of paramagnetic salts in cryogenics, which dates from 1926, and dynamic nuclear orientation, which dates from the 1960s, to achieve, in effect, a "cryogenic cooling" of gravity.

POSTSCRIPTS

Appendix A

Errata from the 1981 Paper

INCORPORATE **the following corrections** into "Anti-Gravity with Present Technology: Implementation and Theoretical Foundation," AIAA/SAE/ASME, 17th Joint Propulsion Conference, 27029 July 1981, Colorado Springs, Colorado. To download, go to Google and search on the terms "Alzofon, anti-gravity, AIAA". The top link will take you directly to the relevant page on the AIAA website. Before you download the PDF, you will have to create an AIAA account and password, which is easy to do. The fee for the download as of this writing is $25.00.

1. Page 1, column 1, paragraph 3, sentence 2

Currently reads:

"These features, which concern the annihilation and creation of matter-energy..."

Should read as follows. Note that the word "energy" changes to "radiation," and a new sentence has been added, beginning "It is emphasized." Italic emphasis by F.A.:

These features, which concern the annihilation and creation of matter-radiation, are shared by all subatomic processes and have been successfully been used in the explanation of the Lamb-Retherford effect, i.e., by the quantum electrodynamics. It is emphasized here, and in the following, creation and annihilation of *both* photons and particles (i.e. radiation *and* matter) are included. The model proposed...

2. Page 3, column 1, paragraph 1, sentence 3

Should read as follows, with an added footnote attached to the word "stability," which directs the reader to *Appendix B* of the paper (and not *Appendix B* of this book):

To remain stable, in the sense in which we have defined stability,* an alteration in the mass-energy distribution must take place, according to Le Châtelier's principle, which will compress the matter-energy into a smaller volume with the smaller diameter given above and corresponding to the new stability configuration.

*See *Appendix B*.

Appendix B, page 31 of the paper, begins with the sentence, "By means of Rayleigh's method, it is possible to gain further insight into the role played by random phase shifts in generating that property of matter-radiation known as inertia."

3. Page 16, column 2, paragraph 2, line 4

Insert equation after "it is clear that one should obtain, for example..." and add harpoon over r, as shown below:

$$<V(\vec{r},t;\vec{r}_0,t_0)>_0 = V(\vec{r},t)$$ where $V(\vec{r},t)$ denotes a solution of Maxwell's equation.

4. Page 16, column 2, paragraph 3, line 5

Add harpoons over the two r symbols, as shown here: $\vec{r}, \ t, \ \vec{r}_0,$ and t_0

5. Page 19, column 2, next-to-last paragraph

Equation 6.42 is missing. The missing equation is shown below ("k" denotes "kappa"):

$$e^{i\,\vec{k}\,\cdot\,\vec{r}} \qquad e^{i\,x\,s_0\,\cos\alpha} \tag{6.42}$$

A Notebook Fragment

EDITOR'S NOTE

The following notebook fragment was probably written between 1989 and 1994, as it briefly compares the UFT with Dr. Hal Puthoff's theory of the origin of gravitation (1989),[74].and the 1994 experiment isn't mentioned Puthoff and my dad were colleagues at SRI in the 1950s and remained in touch on a friendly basis through the 1990s. The topic is the analogy between cryogenics and reduction of the gravitational field by dynamic nuclear orientation. This topic is discussed elsewhere, so it didn't bear inclusion in the main body of the text, but the treatment is somewhat different here, so it seemed worthy of inclusion in an appendix.

INTRODUCTION

It is the purpose of this exposition to compare the cooling of a paramagnetic specimen by adiabatic demagnetization, and the cooling of the gravitational field by dynamic nuclear orientation. These will be compared in detail, with numerical estimates of the quantities introduced, so as to clarify the process of reduction of the gravitational field.

COOLING OF A PARAMAGNETIC SALT BY ADIABATIC DEMAGNETIZATION

1. A constant magnetic field is applied to the specimen, which may have been previously cooled by a convenient means. The field aligns the elementary magnetic dipoles of the specimen, which work on the rest of the specimen, raising its temperature, if applied quickly enough that the heat generated cannot be conducted away quickly enough.

2. The specimen then comes to thermal equilibrium with its surroundings.

3. The magnetic field is removed suddenly, i.e. quickly in comparison with the molecular processes which conduct heat to the specimen, i.e. in a time interval small compared with the time required for a molecule of the environment to traverse one mean free path. The elementary dipoles of the specimen are left oriented.

---

[74] H. E. Puthoff, "Gravity as a Zero-Point Fluctuation Force," Phys. Rev. A 39 [1989] 2333

4. The thermal collisions of the surroundings perturb the orientation of the oriented dipoles, causing them to lose their orientation, and, in the process, losing some of their own energy.

The loss of energy described in Step 4 above is worth discussing in more detail: The surroundings, in equilibrium, is composed of molecules colliding with one another in a chaotic manner with orientations distributed in a random manner. However, when they collide with the oriented molecules of the specimen, some of the orientation of the specimen is communicated to the surroundings, thus removing some of the degrees of freedom available to the surrounding's molecules which have collided with the molecules of the specimen. The rest of the molecules of the surroundings then collide with the oriented molecules and, with the speed of sound, the loss of degrees of freedom is communicated to the rest of the surroundings. This lowers the temperature of the surroundings. This loss of temperature is shared with the specimen.

DYNAMIC NUCLEAR ORIENTATION

In reducing the force of gravitation by dynamic nuclear orientation, the procedure is analogous to the above:

1. A constant magnetic field is applied to the specimen. This field causes the electrons in the specimen to precess with the Larmor frequency (approximately) about the direction of the field. The electrons should be in penetrating orbits so that the precession can be carried to the nuclei of the atoms.

2. A microwave field of the same frequency as that of the precession is applied to the electrons, and this causes the electrons to flip over, and, in turn, causing the nuclei to turn over. Thus the nuclei become oriented.

3. The magnetic moments of the nuclei and electrons are conceived as consisting of two parts:

    a) A hard core of dense, circulating electric currents, essentially responsible for the usually-measured values of the magnetic moments.

    b) A cloud of electric charges due to the virtual processes, which are not as tightly bound to the hard core as the currents making *up* the core.

4. The orientation of the nuclei and electrons induces orientation of the virtual charges as well as the hard core. Since most of the mass of the atoms is due to the nuclei, and virtual processes are responsible for the existence of inertial mass, we can expect there to be more virtual processes near the nuclei than near the electrons. In general, of the mass m, the fraction $\sqrt{G}m$ is effective in producing the gravitational field. Since the field is produced by the overlap of the virtual particle/radiation field attributable to and due to each element of inertial mass, we must attribute

most of the effective mass $G\,m$ to the overlapping field. And since about 80% of the virtual processes in quantum electrodynamics is due to the radiation field, this quantity is approximately equal to the strength of the radiation processes associated with the virtual processes causing the gravitational field.

5. Since the overlap of gravitational fields in space does not alter these fields separately, we must conclude that the virtual, mostly-radiation field generating gravitation must interact with the elementary particle or nuclear masses themselves. The fact that the same fractional amount of this interacting mass: G M, is effective in the force law, is not surprising in light of Puthoff's theory of the origin of the gravitational field (H. E. Puthoff, "Gravity as a Zero-Point Fluctuation Force," *Phys. Rev. A 39* [1989] 2333). This theory is equivalent to the Alzofon theory, except that the approximation is made that only fluctuating dipoles interact through the zero-point electromagnetic field. In terms of the present discussion, we imagine fluctuating dipole moments in a given amount of inertial mass m, generating a fluctuating electromagnetic field, and interacting with the induced fluctuating dipole moments of the mass M. The effect of the fluctuations has been averaged out so that the dipole moment does not appear in explicit form. However, we imagine the dipole moment to consist, essentially, of electron-positron pairs, since, after photons, these take the minimum amount of energy to be created. The distance between these parts of the dipole has been averaged out with the rest of the form factors. Thus, we can expect the total number of particles (or charges) to be

$$\frac{2\sqrt{G}\,m}{e}$$ for a mass m. The magnitude of 2 *G/e* is 1.07 x 10^6 for *m* in *gm*.

[*The fragment ends here*]

Appendix C

Publications on Gravity and Gravity Control

THE FOLLOWING list includes two reviews of Alzofon's theory and his research on gravitation. His work on gravitation was mentioned briefly in *OMNI* and *Science Digest*, probably in the 1980s, but exact references are unavailable.

1a. "The Origin of the Gravitational Field," *Advances in the Astronautical Sciences*, Vol. V, pp. 309-319, Plenum Press, New York, 1960. [*Ed note: Incorporate errata from* Appendix A *above.*]

   This is the paper that drew the attention of Air Force analyst Dr. Maurice Garbell, who was conducting a worldwide survey of gravitation research (see Chapter 12, p. 71). In the report, Garbell stated that the UFT suggested a means of controlling gravitation. In a conversation at a restaurant in Palo Alto, he said it was the only theory of gravitation he'd found that had a prayer of becoming an applied technology, but at the same time, he predicted that academia would never investigate it. His prediction has held true for 56 years. Garbell's paper is significant enough to cite below.

1b. M.A. Garbell, "Soviet Research on Gravitation, An Analysis of Published Literature," sponsored by Science and Technology Section, Air Information Division, AID Report 60-61, 379 pages. Distributed by U.S. Department of Commerce, Business and Defense Services Administration, Office of Technical Services, Washington, D.C., October 1960. *Ed note: Available online.*

2. "Anti-Gravity With Present Technology: Implementation and Theoretical Foundation," AIAA/SAE/ASME, 17th Joint Propulsion Conference, 27029 July 1981, Colorado Springs, Colorado

   In order to download a PDF, go to Google and search on the terms "Alzofon, anti-gravity, AIAA". The top link will take you directly to the relevant page on the AIAA website. Before you download the paper, you will have to create an AIAA account and password, which is easy to do. As of this writing, the fee for the download is $25.00. Be sure to incorporate the errata in *Appendix A*, p. 253.

3. "A UFO Propulsion Model," MUFON Symposium Proceedings, *MUFON Journal*, No. 170, April 1982

4. "The Unity of Nature and the Search for a Unified Field Theory," *Physics Essays, volume 6, number* 4 (1993) 599-608

    This paper complements the theoretical foundation in Ref. 2 above. My father considered it his most important paper on the UFT. Of particular interest is Sec. 7, Model of Gravitation and Alteration of the Gravitational Force

5. Yost, Charles; "Review of the Work of Dr. Frederick E. Alzofon," *Electric Spacecraft Journal*, compiled and reviewed by Editor Charles Yost, Issue 13, 1994

    The late Charles Yost conducted a telephone interview with my father and wrote this comprehensive statement of the theory and technology. My dad later said that while it was a good introduction, particularly for lay readers, it was inaccurate on several points. The article mentions the success of the 1994 experiments. Unfortunately, copies of the Journal are extremely difficult to find.

6. Fox, Hal, Editorial: "New Science Teaches, Soon Engineers Build," *New Energy News – Monthly Newsletter of the Institute for New Energy*, Vol. 2, No. 7, ISSN 1075-0045, Nov. 1994. PDF available online at http://newenergytimes.com/v2/archives/fic/N/N199411s.PDF as of this writing.

    Fox reviews the unified field theory in one paragraph, and in the final two sentences says, without equivocation, that my father's experiments successfully substantiated the theory. It is unknown where Fox got his information, since no source was cited, although it is possible it was Yost (5 above), who also alluded to successful experiments. At any rate, the cat was out of the bag at this point, and no-one seems to have noticed.

7. "Light Signals, the Special Theory of Relativity and Reality," *Physics Essays, 14* (2001) 144-148

8. "UFOs and Crop Circles, Gravitational Field Common to Both," *MUFON Journal*, No. 435, July 2004: https://issuu.com/disclosureproject/docs/mufon_ufo_journal_-_2004_7._july, or search on "Alzofon MUFON Disclosure Project" and follow the top link.

9. "A 'new and simple idea,' dark matter-energy and the crisis in physical theory": This paper is included here. See *Chapter 38*, p. 229.

    In order to download original, go to http://vixra.org/pdf/1007.0008v1.pdf

    This was my father's final paper on the UFT, written when he was ninety years old. The paper was turned down by *Nature* and *Physical Review*, which he felt was due to the same old prejudices that dated back to the 1940s. While there is factual evidence for this, this is not the proper venue to elaborate. He told me he didn't have the stomach for another time-consuming battle with editors and referees, so he filed the paper online at ViXra.org in 2009 in order to establish a copyright on the text and ideas. No permissions were required to reprint it here.

The following is a copy-and-paste of text from the ViXra.org website:

ViXra.org is an e-print archive set up as an alternative to the popular arXiv.org service owned by Cornell University. It has been founded by scientists who find they are unable to submit their articles to arXiv.org because of Cornell University's policy of endorsements and moderation designed to filter out e-prints that they consider inappropriate. ViXra is an open repository for new scientific articles. It does not endorse e-prints accepted on its website, neither does it review them against criteria such as correctness or author's credentials.

Appendix D
Publications by F. Alzofon

1. "Relativistic Neutron-Proton Scattering in the Born Approximation," *Physical Review*, 75 (1949) 1773

2. "The Probability of Continuous Tracking in the Presence of Random Noise," *Proc. National Electronics Conf.* 14 (1958) 519

3. "Theoretical and Experimental Investigation of Heat Conduction in Air, Including the Effects of Oxygen Dissociation," NASA Technical Report R-27

4. "An Exchange Transfusion Formula," *AMA Journal of Diseases of Children*, 98 (1959) 694

5. "The Origin of the Gravitational Field," *Advances in the Astronautical Sciences*, Vol. V, pp. 309-319, Plenum Press, New York, 1960

6. "The Effect of Tubing Dead Space in Exchange Transfusions," AMA Journal of Diseases of Children, *102* (1961) 194

7. "Applications of Infrared Technology in Nondestructive Testing," *Proc. Missiles and Rockets Sym.*, USNAD, Concord, CA, 18-21 April 1961, pp. 247-251, Fontes Abbey Press, 777 W. Grand Avenue, Oakland, CA

8. "Retardation and Diffraction Aspects of the Conduction of Heat in Solids," *American Journal of Physics, 30* (1962)

9. "Infrared Nondestructive Testing of Glass Filament Wound Rocket Motor Cases," ASTM Fourth Pacific Area National Meeting, Proc. 1962

10. "Variational Approach to the Calculation of Charge Exchange Cross Sections for Adiabatic Collisions." NASA TN D-1654, May 1963

11. "Factors Influencing the Detection of Flaws in Glass Filament Wound Rocket Motor Cases by

Infrared Scanning," Second ICPRG Symposium on the Nondestructive Inspection of Solid Propellant Rocket Motors, Chemical Propellant Information Agency, Applied Physics Laboratory, Johns Hopkins University, 1963, Proc.

12. "An Infrared Nondestructive Testing System for Rocket Motors," *Materials Evaluation, 23* (1965) 537

13. "Relative Contributions of Emissivity and Thermal Conductivity in Infrared Nondestructive Testing," Transactions of the Infrared Sessions, Society for Nondestructive Testing, Spring Meeting, 22-26 February 1965, Los Angeles, CA, Proc.

14. "Optics and Infrared Nondestructive Testing," Symposium on Physics and Nondestructive Testing, 28-30 September 1965, Dayton, Ohio, Proc.

15. "Infrared Detection of Flaws in Adhesive-Bonded Metallic Honeycomb," Society for Nondestructive Testing, 1966 Spring Convention, 7-10 March 1966, Los Angeles, CA, Proc.

16. "Infrared Evaluation of Microweld Quality," *Materials Evaluation, 25* (1967) 183

17. *Multiple-Valued Functions in Three-Dimensional Space and Sommerfeld's Method*, published under the auspices of the Houston Aerospace Division, Lockheed Electronics Company, 1970. (Book)

18. "The Optical Environment of Rocket Platforms," Proc. Infrared Information Symposium, *14* (1970) 131 (This paper was classified CONFIDENTIAL when published, but has since been declassified)

19. Many reports on infrared technology have been written but are not cited here. Some of their titles are quoted in the *Bibliography of Practical Applications of Infrared Techniques*, by Riccardo Vanzette, John Wiley and Sons, New York, 1972

20. "Remote Sensing of Thermal Subsurface Terrain Properties," Technical Papers of the American Society of Photogrammetry, 47th Annual Meeting, Washington, D.C., 22-27 February 1981, pp. 335-341

21. "Anti-Gravity With Present Technology: Implementation and Theoretical Foundation," AIAA/SAE/ASME, 17th Joint Propulsion Conference, 27029 July 1981, Colorado Springs, Colorado (see p. 259 for download information)

22. "Comparison of the Day and Night Signatures of a Strategic Target" (U), SECRET, Infrared

Information Symposium, 14 February 1985, Orlando, Florida

23. "Exact Solutions in Closed Form for Fresnel Diffraction by a Semi-Infinite Plane and a Circular Disc," Proc. SPIE, Current Developments in Optical Engineering and Diffraction Phenomena, *679* (1986) 112

24. "Diffraction of a Plane Electromagnetic Wave by a Flat Conducting Strip and by a Circular Conducting Channel," Proc. SPIE, Current Developments in Optical Engineering and Diffraction Phenomena, *679* (1986) 119

25. "The Optics of Heat Conduction in Solids," Proc. SPIE, Technical Symposium Southeast, Thermosense IX, *780* (1987) 215

26. "Scattering of Electromagnetic Radiation by a Hexagonal Ice Crystal," Proc. 14th Congress of the International Commission for Optics, 24-28 August 1987, Quebec, Canada, pp. 405-406

27. "Radiation Scattering by Flakes of Arbitrary Shape," Proc. SPIE Technical Symposium Southeast, 4-8 April, 1988, Orlando, Florida, 927 (1988) 10

28. "Radiation Scattering by Solids of Arbitrary Shape: A New Method of Solving Boundary Value Problems," *Journal of Wave-Material Interaction*, *3* (1988) 85

29. "Scattering of Traveling Temperature Waves by Small Inclusions in a Homogeneous Solid," *Journal of Wave-Material Interaction*, *3* (1988) 283

30. "Multiple-Valued Solutions of the Equations of Field Propagation: Sommerfeld's Method, I, Introduction," *Journal of Wave-Material Interaction*, *7* (1992) 121

31. "Multiple-Valued Solutions of the Equations of Field Propagation: Sommerfeld's Method, II, Static Fields," *Journal of Wave-Material Interaction*, *7* (1992) 127

32. "Multiple-Valued Solutions of the Equations of Field Propagation: Sommerfeld's Method, III, Fraunhofer Diffraction," *Journal of Wave Material Interaction*, *8* (1993) 1

33. "Multiple-Valued Solutions of the Equations of Field Propagation: Sommerfeld's Method, IV, Fresnel Diffraction," *Journal of Wave-Material Interaction*, *8* (1993) 89

34. "The Unity of Nature and the Search for a Unified Field Theory," *Physics Essays*, *6* (1993) 599-608

35. "Optical Analogues in a Dispersive Medium, I, The Physical Model and Geometrical Optics," *Journal of Wave-Material Interaction*, *8* (1993) 185-199

36. "Optical Analogues in a Dispersive Medium, II, Physical Optics and Anisotropic Mediums," *Journal of Wave-Material Interaction*, *8* (1993) 200-217

37. "Optical Analogues in a Dispersive Medium, I and II, have been reprinted in *Selected Papers on Temperature Sensing: Optical Methods*, Ed. R. D. Lucier, SPIE Milestone Series, Vol. MS 116, SPIE Optical Engineering Press, Bellingham, Washington, 1995

38. "Some Optical Concepts in the Analysis of Viscous Fluid Flow," *Journal of Wave-Material Interaction*, *11* (1996) 219-233

39. "The Transition to Turbulent Fluid Flow," *Journal of Wave-Material Interaction*, *11* (1996) 234-244

40. "Light Signals, the Special Theory of Relativity and Reality," *Physics Essays*, *14* (2001) 144-148

41. *Two Methods for the Exact Solution of Diffraction Problems*, SPIE Press, Bellingham, Washington, 2004. (Book)

42. "A 'new and simple idea,' dark matter-energy and the crisis in physical theory": Final completed paper, unpublished, filed online in 2009 (age 90), at the following URL:

http://vixra.org/pdf/1007.0008v1.pdf

From 2010 to 2012, when he was 91 to 93 years old, my dad was working on a further extension of Sommerfeld's Method. After he died, circumstances allowed only two hours to sort through the reams of papers in his study, and I was unable to find any drafts of the new book.

Appendix E

Degrees and Professional Associations

Ph.D., Mathematics, 1956, University of California, Berkeley. Dissertation: *Multiple-Valued Functions and Sommerfeld's Method*, under Prof. Griffith C. Evans.

M.A., Physics, 1948, University of California, Berkeley.

A.B., Mathematics (minor in Physics), 1941, University of California at Los Angeles.

Honors: Publication Incentive Award, Lockheed Missiles and Space Company, June 1963.

Societies: American Association for the Advancement of Science; Optical Society of America; American Association of Physics Teachers; Society of the Sigma Xi, Pi Mu Epsilon; American Mathematical Society; Society for Industrial and Applied Mathematics; Research Society of America; Institute of Electrical and Electronic Engineers (IEEE); Member of the Thermal Committee; Member of the Thermal Committee of the Society for Nondestructive Testing

Additional coursework: Completion of night school courses in Palo Alto, California: Television Repair, 33 hours, 1/58 to 3/58; Electronics III, 24.5 hours, 4/58 to 6/58; Electronics IV, 42 hours, 8/58 to 12/58; Electronics V, 31.5 hours, 1/59 to 3/59; Electronics VI, 21 hours, 3/59 to 6/59; Electronics Laboratory Workshop, 21 hours, 7/59 to 8/59; Lockheed Missiles and Space Company Course, 20 hours, Fortran Programming, completed 1960. Also many courses in Management, Accounting, Communication, and Machine Shop Practice.

Appendix F

Patent Information

ON the one hand, this book seems to offer the "secrets" of gravity control freely and without restriction. On the other, the DISCLAIMER (p. 5) states the following:

Dr. Alzofon filed a patent application for the technology described in this book in 1980. The patent application was turned down and went into suspension afterward. ***PUBLICATION OF THIS BOOK DOES NOT CONSTITUTE A SURRENDER OF DR. ALZOFON'S PRIOR PATENT CLAIM***, *especially since we believe ("we" being Dr. Alzofon up until the time of his death, his patent attorney, other experts, and the Editor of this book) that the application was treated prejudicially and denied for insufficient cause.*

The cause given was the lack of a working device. However, as my father's patent attorney pointed out, lack of a working device hasn't stopped the patent office from granting patents in the past, as long as the components of the technology are based on known physical processes. This may explain some of the dubious "Star Trek" patents now on file.[75] If these patents represented real technology, the world would have changed, and it *hasn't*—twenty, thirty, *sixty* years after the claims were made public.

For some reason, the patent office raised the bar for my father's application. If and when someone successfully duplicates the 1994 experiment, they might like to join forces with me to revive the 1980 patent filing (see back page for contact information). The reward would be a more than generous division of revenue in favor of the party who provided the working device. If someone wants to file a patent application *without* contacting the author, they are free to do so. They would be relying on the notion of "prior art," that is, the exposure of the technology to the public eye since 1981. However, the same doctrine would also *sink* their application, because *no patent can be filed on prior art*. Once proof of concept emerges, one can imagine the army of lawyers who would volunteer their services to the cause of mounting lawsuits on the basis that the patent should have been granted in 1980. I would be inclined to listen to them, but if those suits fail, the publication of the 1981 paper may turn out to be the best thing for the future of the planet, since it would allow the technology to spread unimpeded.

As far as patents go, the two best courses of action would be to join forces with the author as described above, or to file a patent on an *actual* gravity-control air car or similar device. One can imagine that

---

[75] Four of these are reported in *The Secret History of Extraterrestrials*, by Leo Kasten, (Bear and Company Books, Rochester, VT, 2010, p. 164 – 167), but this is by no means an exhaustive list of "patents without devices."

at least a hundred thousand patents could be generated in this manner (see p. 19). Think, for example, of all the patents generated by the internal combustion engine. The parallels are clear enough. However, working devices would be required.

If I had allowed concerns about patents and money to contaminate my motives for publishing this book, it never would have been written. As stated here repeatedly, the technology must spread far and wide and quickly—*very* quickly—if its benefits are to be experienced in time. Money means delay, as I have so often seen in Silicon Valley. Whenever money sat down at the table, insanity and bloodletting began. One need look no further than the current state of our beautiful blue planet, which has been carved up, poisoned, and damned near murdered—all for the love of money. As the Romans said, "*Radix malorum est cupiditas,*"[76] and they were translating from the Greeks, who were translating from the Bible. The problem of greed has been around for a long, long time, in other words. I can't think of a way of changing it. This book represents a different strategy: *enlisting* greed rather than *resisting* it. The result: a lifting of bureaucratic barriers and opposition from the fossil fuel industry, resulting in rapid, worldwide dispersal of a green technology. Folks, you can *make money* on this stuff— assuming it works, of course.

Quotes appeared around the word "secrets" above That's because there *are* no secrets in this book. Everything's been out in the open since 1981. Anyone with the wherewithal and the expertise could have built a gravity control device, simply by gathering my father's publications and familiarizing themselves with the technology. It's all there: the process, the configuration of elements, the materials used, the microwave frequency, the pulse rate, the electromagnetic field strength, and all the source material from which these figures were derived.

Given these facts, I do not hold out much hope for reviving the 1980 patent application. However, in addition to the humanitarian purposes that inspired this book, it was important to establish my father as the inventor, should anyone else try to take credit for gravity control if and when it is fully proven. It was also important to put the theoretical basis for the technology on the record—yet again—so that physics can advance into the twenty-first century with the right tools for the job at hand: the conquest of space and our graduation to cosmic consciousness, if not cosmic citizenship. In this, I can only hope, the present volume has succeeded.

---

[76] "Greed [or 'the love of money'] is the root of all evil."

AUTHOR BIOGRAPHY

The Inventor, a Brief Life

FREDERICK ALZOFON was born in Detroit, Michigan, in 1919, the youngest of six children, including four sisters and a brother. As a scientist, he never gave any credence to astrology, but, as a child of the '60s, I can't resist noting that at the time of his birth virtually all the planets were in one house, implying enormous concentration in one area. As it happened, the chart proved to be prophetic, as he was to concentrate all of his considerable mental resources on physics throughout his life, devoting himself in particular to the search for a unified field theory that would resolve the problems of quantum mechanics and general relativity.

If you are old enough to remember Lt. Columbo, the TV detective in the "dirty raincoat" who bedeviled upper-crust murder suspects with "one more question," you have a pretty good idea of the way he carried himself. He walked with a rapid, deliberate gait, slightly bent over at the waist, absorbed in his own thoughts, which ran along a different track from everyone else and were generally rather serious.

Like Columbo, he wore the same threadbare, beige raincoat, white shirt and blue tie to work, day after day, no matter what the weather, for decades. Like Columbo, he was a bit goofy: He drove a clunky old beater (a 1950 Plymouth with a stick shift), and paid no attention whatsoever to traffic. A CHP officer pulled him over on the Bayshore Freeway one morning in the 1960s and asked him why he was going forty miles an hour when the speed limit was sixty-five. "I was thinking about equations," he said. It was a novel excuse, but it didn't get him off. I feared a car ride with my dad more than a ride on the most radical roller coaster in creation.

As far as his manner of speaking, the only public figure with whom I can find any parallels is Noam Chomsky. Like Professor Chomsky, he seemed to know everything about everything, and he had processed it all into some new and higher form of knowledge. He was an extraordinarily fast reader and retained everything he ever read. As far as literature, he liked Charles Dickens and Shakespeare better than the Russian or French novelists. I can't say what he liked in modern literature, but whenever I mentioned anything, he had read it. Case in point: Lee Child. When I caught up on all of the Reacher novels, I told him he might enjoy them, and it turned out he'd already read every single one, and I hadn't had a clue. He was an avid reader of science-fiction from the 1930s through 2012. While he was devoted to physics, he was a keen observer of politics, pop culture, social movements, and the cinema. If you asked a question of him, you were likely to get a lecture, but it would be a pleasant one, incredibly dense with information, rife with unexpected insights, with long arcs of reasoning navigated in a calm voice, as if read from the pages of a book—again, much like Professor Chomsky.

Politically? *Not* like Chomsky. Not like anyone else in the world, either. Ever. Another book will be required, and this is not the place to begin writing it.

His parents were Russian Jews who fled Odessa in the early 1900s, after sword-wielding Cossacks began riding through the streets, slashing random villagers.[77] Friends assured them that the violence was a temporary aberration and would soon blow over. The "temporary aberration" turned out to be a lethal tide, however, and the villages they left behind, Duvassar and Hirhutsk, return no results on Google today—obliterated by two world wars, in all likelihood. My father grew up in a rough and tumble part of Detroit the locals called "Hell's Kitchen." In an odd parallel, boxer Joe Louis grew up in the same area at the same time after *his* parents fled racist persecution in Alabama in 1926.

Because of his humble origins, my dad developed a lasting empathy for oppressed people and humanitarian causes, but he was especially sympathetic toward animals. My first memory is of my dad trying to save the life of a tiny chick that had fallen out of its nest and was being pecked to death by blue jays. This occurred on Point Loma in 1954, when I was four. When we lived in Santa Barbara I remember him jumping in the water at the marina to save a drowning dog. The dog showed his gratitude by drenching my dad in a shower of salty spray as they stood side-by-side on the dock after the dramatic rescue. Toward the end of his life Frederick was contributing regularly to the Jane Goodall Institute.

He was singularly unimpressed by fashion, social status, or the trappings of wealth and power. The expression "marched to the beat of his own drummer" was probably made up with him in mind. This attitude often put him at odds with authority. He didn't like conflict; he simply insisted on calling the truth as he saw it. Period.

Philosophically, he called himself a logical positivist. This isn't surprising, since Professor Lenzen was a student of Bertrand Russell. Intellectually, his greatest gift was an ability to absorb enormous

---

[77] My granddad on my mom's side was from the Frisian islands, off the coast of Denmark. He was a pipe-smoking sailor who stubbornly insisted on navigating the highways of Florida and Ohio by the sun and stars. Needless to say, he got lost a lot. My grandmother came from a village on the border between Scotland and England. Her nationality depended on who was winning the war at the time. She was a redhead, however, so she probably leaned toward Scots. My mom had Nordic features. She was tall, kindhearted and compassionate, with blue eyes and blonde hair, but she was definitely an earthling, not a Pleiadian.

amounts of information and see abstract patterns within all the confusion that others did not see. He also had an unmatched talent for generalization, that is, inductive reasoning. He viewed science as an alternation between inductive and deductive reasoning. I often asked him how he acquired his talent for generalization, and he said, "Practice." I didn't believe him. Practice may have sharpened it, but it was a natural talent and quite off the scale.

Though his dad was a watchmaker and a gunsmith, Frederick became fascinated with science, particularly physics and mathematics, while a preteen, and he began self-teaching by reading all the books he could find on these subjects in the city library. The stories of the ancient Greek mathematicians, Archimedes and Euclid, had a profound effect on him, and I remember him lecturing me on the origins of "Eureka!" when I was six years old. The triumphs of modern physics, especially the genius of Albert Einstein, inspired him. When he graduated from high school he was offered a music conservatory scholarship but declined in order to pursue a career in physics and mathematics. Ironically, he began losing his hearing in his thirties and was almost deaf by the time he was seventy.

The path of science led him to UCLA. In 1941, he graduated with an A.B. in Mathematics and a minor in Physics. As an undergrad he worked in a prop shop on Sunset Boulevard in Hollywood and carried a spear as an extra in a production of *Aida*.

In 1942 he entered the graduate program in physics at Cal Berkeley, where he conducted research in elementary particle collisions at the cyclotron (E.O. Lawrence, who received a Nobel for inventing the cyclotron in 1939, was head of the laboratory at the time). His professors included J. Robert Oppenheimer, David Bohm (who was working on his PhD with Oppenheimer as his advisor), J. W. Weinberg, Victor Lenzen, and Raymond Birge. Nobel Prize-winning chemist Robert Millikan sat on his doctoral committee. First-hand anecdotes about these larger-than-life figures were a steady staple of my childhood.

Frederick received an M.A. in Physics in 1948, and pursued a doctorate as far as his PhD orals, but he abruptly switched to the Department of Mathematics, even though it set him back years. In 1956 he received his doctorate under Professor Griffith Conrad Evans (1887 – 1973), Department Chairman and namesake of Evans Hall of Mathematics. It is said that he was Evans' last PhD student. When my dad received his doctorate, he called his mother to celebrate, and she told him that he was the wrong kind of doctor.

Apparently his break with the Physics Department came about when, after he had already passed his orals, a hostile professor ambushed him with an additional exam over and above what was required. The conflict may have been exacerbated by my dad's tendency to correct his professors' blackboard equations in front of the class (something I learned about from my mom). His focus was particle physics, but the added question was something like the following: "You have a twenty-foot wire with a short circuit somewhere on it. How do you locate the short circuit?" When he couldn't answer the question, they failed him. Later, he found out that the exam had been optional, but they hadn't told him. That's when he abandoned physics and switched to mathematics. While he had nothing but praise for Professor Evans, the episode in the Physics Department made a lasting impression on him and was partially responsible for his decision to leave academia for aerospace in the early 1950s, and his distrust of academia before, during, and after the 1994 experiment.

His doctoral dissertation was entitled *Multiple-Valued Functions and Sommerfeld's Method*. The virtue of Sommerfeld's Method was that it provided exact solutions to diffraction problems. Its limitation was that it could only be applied to an infinite half plane. Expert opinion, including Sommerfeld's, was that the method could not be extended to objects of arbitrary shape. In 2004, when he was eighty-five years old, he published *Two Methods for the Exact Solution of Diffraction Problems* (SPIE Press, Bellingham, Washington), a theoretical treatise that succeeded in extending Sommerfeld's Method to objects of arbitrary shape. The book, which has implications for DVD technology and radar imaging, has been the subject of research at the University of Taiwan. One of the professors visited him in Oregon. Bootleg copies appear in technical libraries all over China, a fact that amused my dad. He always felt that Sommerfeld's Method had a wealth of unrealized potential, and he was working on another book on it at the time of his death in 2012.

While in the PhD program at the University of California in the 1940s, he took a keen interest in relativity because, among other things, it had a direct bearing on his research at the cyclotron. In 1949, he published "Relativistic Neutron-Proton Scattering in the Born Approximation" in *Physical Review* (75, 1949 1773). He was also having frequent discussions with Professor Victor Lenzen (1890 – 1975), an acknowledged expert on the subject. It was during this period that he conceived of a unified field theory based on a minor revision of special relativity.

The mounting responsibilities of marriage and children caused him to leave academia for aerospace research in the early 1950s. From 1951 to 1952 he was a researcher at the Naval Radiological Defense Laboratory at Hunters Point in San Francisco. From 1952 to 1953, he conducted research in long-range underwater sound transmission at the Navy Electronics Laboratory in San Diego and taught calculus to artillery specialists. My earliest memories date from the period we lived in a drafty cabin in a eucalyptus grove on the Pacific side of Point Loma and later when we lived in the Sunset Cliffs neighborhood. From 1953 to 1956, he was a physicist at the Santa Barbara Research Center in Goleta, where he conducted research in the theory of infrared detector technology.

Though he had become disaffected with academia, his love of research never wavered, and, in spite of having a full plate at work and at home, he continued to invest every spare moment in his unified field theory. After a meeting with Richard Feynman at Caltech in 1954, he dedicated all of his spare time to publishing his theory of the origins of gravitation (see *Chapters 10* and *11,* pp. 59 - 70). This was in addition to work on his PhD thesis.

In 1956, he was hired as a mathematician at SRI (Stanford Research International) and moved to Palo Alto. There he met Professor Hal Puthoff of Stanford, whose ideas on gravitation paralleled his own. He and Puthoff became friends and remained in touch on a collegial basis at least through the late 1990s. According to Google, Dr. Puthoff is now CEO of the privately funded research organization, Earthtech International, Inc., in Austin, Texas. As gravity control is one of the prime objectives of the organization, it is hoped that this book will light the way toward success in this area for them.

In 1960, he published "The Origin of the Gravitational Field" in *Advances in the Astronautical Sciences*, a publication of The American Astronautical Society (Vol. 5, Plenum Press, NY; see p. 71 for more).

From 1958 through 1969 he was a staff scientist at Lockheed Missiles and Space Company, Sunnyvale. Among his accomplishments there were experiments he conducted proving the feasibility of infrared nondestructive testing, and moreover, its superiority to X-rays in the detection of structural flaws in rockets or soft materials, such as O-rings. When it became clear that funding cutbacks were coming, he transferred to the Houston Aerospace Division of Lockheed Electronics Company, where he supervised analysis of data acquired by optical instruments.

While working at Lockheed, Houston, he completed his first book, *Multiple-Valued Functions in Three-Dimensional Space and Sommerfeld's Method*, which was published under the auspices of the Houston Aerospace Division of Lockheed. The origin of the book, he told me, was his observation of anomalies in collapsing magnetic fields while doing research at the cyclotron. He found it convenient to explain these anomalies in terms of multi-leaved spaces, or "multiple universes." You would never know it to read the mathematically-oriented book, but he always felt that the multiple universes he described might explain ESP or what people referred to as "the afterlife." He wanted to devise experiments to test this hypothesis, but he never had the opportunity to follow up.

In the summer of 1970, while I was living in Houston, my parents agreed to a divorce. In August, Hurricane Celia, a Category 3 storm with winds of 125 mph, hit the Gulf Coast, and foolishly we decided to weather it out in Seabrook, less than a mile from the water's edge. On the day the storm made landfall, the sun set at three o'clock in the afternoon. The rising Gulf inundated the exit roads while we boarded up our windows. That night, the storm descended in full fury. The winds nearly ripped the house apart while the deluge hammered walls and windows with terrifying force. In the street, a barrage of lightning bolts struck everywhere all at once—houses, telephone poles, trees, drainage gratings, and the lightning rod on our roof—leaving us cringing like soldiers in a trench with artillery shells exploding all around. Perhaps this was the origin of my respect for climate change.

From 1976 through 1978, my father moved to Long Beach to work as an infrared and optics engineer at Rockwell International, where he analyzed the performance of satellite-mounted military observation systems.

In 1978 he moved to Seattle to become a senior staff scientist at Boeing Aerospace, again working in infrared and optics design. He repeatedly declined promotions in order to continue his spare-time research on Sommerfeld, gravitation, and unified field theory. This led to his 1981 paper, "Anti-Gravity with Present Technology: Implementation and Theoretical Foundation," which was delivered at the 17th Joint Propulsion Conference in Colorado Springs (see *Chapter 14*, p. 81). Convinced of the value of researching the proposed technology, he began to seek financial backing from government, private industry, academia, and research foundations.

In 1984, at the age of 65, he retired from Boeing and moved to Corvallis, Oregon. It was the beginning of the most creative period of his life. Finally freed from the necessity of working for others, he devoted himself entirely to research on gravitation and other topics, including heat conduction in solids (to which he applied techniques of optical analysis), the transition to turbulent flow in fluids (a problem that had defied complete understanding and mathematical expression for a hundred years), and of course, Sommerfeld's Method. All of his papers and books on these topics were published. In 1994, he finally succeeded in mounting experiments to test his theory of gravitation. The experiments were a success, but he decided to keep the results a secret while he sought investor backing.

In 2003, at the age of 84, he delivered an address on the special theory of relativity and the speed of light to the American Association of Physics Teachers in Guelph, Canada, and received a standing ovation. By then he had become tired of the effort to find backers, so he returned to fulltime research in Sommerfeld's Method and unified field theory. Even in his nineties he made daily trips to the library at the nearby University of Oregon to order foreign-language physics journals and rare publications, such as a book by maverick physicist Burkhard Heim (1925 – 2001), which was only available in German. He had been interested in Heim since the early 1980s, when he wrote to the theoretical physicist, asking for a copy of his book. The letter had gone unanswered. Part of the reason, one might surmise, is that Heim had lost both hands and most of his hearing and eyesight at the age of nineteen in an explosion at a chemical lab in Nazi Germany. He was also known to be eccentric and reclusive. What intrigued my father most was Heim's claim to have calculated the mass of elementary particles. After studying Heim's abstruse theory for more than two years, he concluded with great disappointment that its complexity masked a lack of substance.

In 2009, he turned once again to unified field theory and wrote his final paper on the subject, "A 'new and simple idea,' dark matter-energy and the crisis in physical theory," which he dedicated to Richard Feynman (see p. 229).

Beginning in the early 1970s, my dad had suffered from recurring attacks of melanoma. The first led to emergency surgery, during which the tumor was traced deep into his sinus cavity. Part of his upper lip and sinus cavity were removed. In the mid-1970s, while he was living in Long Beach, he developed thyroid cancer and it, too, was removed. In 2012, at age 93, he was admitted to the hospital for weight loss and an inability to swallow. Doctors discovered that a melanoma, which had been caught and excised early, had spread to his right lung. The tumor was inoperable. Upon returning home for hospice care, suffering from starvation and dehydration, his mind clouded by heavy doses of morphine, he sat up in bed and read the *New Yorker* magazine, as he had done routinely for decades. Then he put it down on the nightstand and slipped into a coma. He died peacefully early on a snowy Tuesday in December, 2012.

Doctors who examined my dad for military service during World War II discovered that he had a cardiac arrhythmia and an enlarged heart. They classified him "4F" and told him he would be lucky to live beyond the age of 30. It seems thematic for a man with a lifelong resistance to the pronouncements of authority that he survived until the age of 93 and kept his wits about him until the very end. As someone who knew him, I often think that his ever-active imagination and brilliant intellect were what sustained his body three times longer than the allotment granted by medical science. The irony would not be lost on him: *He defied science to lead a life dedicated science.*

The inscription he chose for his gravestone read simply, "Frederick Alzofon, Scientist." Whatever the fate of the UFT or gravity control, I hope that new generations of scientists will find inspiration in the creativity, dedication, intellectual honesty and rigor, and the endless capacity for hard work evidenced in his unique contributions to mathematical physics and applied science.

About the Editor

David Alzofon, seen here toasting the future of gravity control near Mt. Palomar Observatory in 2016, was born in Berkeley, California, in 1950, attended public school in the San Francisco Bay Area, and graduated from the University of California, Santa Barbara, in 1972, receiving a BFA. He returned to the Bay Area soon afterward and began writing for a local newspaper. In the mid-1980s, he joined the personal computer revolution as a technical writer.

Twenty years, a hundred manuals, and six start-up companies later, Mr. Alzofon decided that time was running short. In 2007, he gave up on Silicon Valley and dot-com dreams and vowed to devote his remaining time to writing only about things that mattered to him.

The thirty-five years between 1972 and 2007 might be considered a mere prelude to the quest that began when his father died in December, 2012, a quest that led to the two-volume set: *Gravity Control with Present Technology*, and *The Top-Ten UFO Riddles*.

Mr. Alzofon currently resides in southern California, where—when not fleeing wildfires spawned by the drought (an early gift of global climate change)—he spends most of the day writing. Currently he has three books planned, one fiction, two nonfiction. The current volume will be the last on his father's technology, but correspondence is welcome (see page 282).

CONTACT INFORMATION

Email: info@klafraknar.com. Include verifiable name, address, phone number, and title and company or educational institution, if any.

Facebook: Dr. Frederick Alzofon

YouTube channel: Alzofonphysics

Copies of the book are available on Amazon.com.